DEDICATION

This book is dedicated to the deeds and exploits, to the sacrifices and achievements of the generation of Australians who proudly served in the RAN's Sea Fury, Firefly and Sea Venom Squadrons.

Y,

AND

OM

ERVICE

n

ACKNOWLEDGEMENTS

Sea Fury, Firefly and Sea Venom is the eleventh in our successful and ongoing *Australian Airpower* series and represents the first one covering naval aircraft exclusively.

Once again, many people need to be thanked for their assistance, for without them it would be impossible to produce the books at the frequency at which they are published whilst maintaining quality.

As always, the people at Aerospace Publications must come first for their continued dedication and enthusiasm: publishers Jim Thorn and Margaret Thorn, production manager Maria Davey and Gerard Frawley, Lee-Anne Simm and Gayla Wilson.

For this book I wish to particularly thank Ross Gillet (RAN PR), Norman Lee and Steve Long plus Harry Margulies for his artwork, John Allen for his drawings and Matthew Tesch, who drew the map of Korea.

Others whose contributions of photographs and other material must be acknowledged include David Wilson, Keith Myers, Bruce Chapman, Mike Kerr, Joe Barr and Richard Hourigan, most of whom have been involved since the first book in the series. In addition, I would like to thank the others whose photographic contributions are noted throughout the book.

Finally, a very special word of thanks to those friends, colleagues and family members who helped me through a very difficult time. Without them, this book wouldn't have been written at all. *"Je veux un mouton qui vive longtemps ..."*

STEWART WILSON
Queanbeyan, 1993.

Published by Aerospace Publications Pty Ltd (ACN: 001 570 458), PO Box 3105, Weston Creek, ACT 2611, publishers of monthly Australian Aviation magazine.
Production Manager: Maria Davey

ISBN 1 875671 05 6

Proudly printed in Australia by Pirie Printers Pty Ltd, 140 Gladstone Street, Fyshwick 2609
Distributed by NDD, 150 Bourke Rd, Alexandria 2015
Film by Pirie Printers Pty Ltd, Canberra
Distribution in North America by Motorbooks International, 729 Prospect Ave, Osceola, Wisconsin 54020, USA. Fax: 715 294 4448

CONTENTS

FRONT COVER: Fairey Firefly 5 VT500/201K of 817 Squadron aboard HMAS *Sydney,* mid 1950s; Hawker Sea Fury FB.11 WH589/115 NW of 724 Squadron based at Nowra; De Havilland Sea Venom FAW.53 WZ930/804M of 805 Squadron aboard HMAS *Melbourne* 1962. (Artwork Z Margulies)

ROYAL AUSTRALIAN NAVY

INTRODUCTION

The Royal Australian Navy Fleet Air Arm suffered a difficult birth in that it took many years of trying and official discouragement before its formation was finally approved in 1947.

Within four years of that, the FAA was proving itself in action in Korea, and two of the three aircraft covered in this book were involved.

The book looks at the formation of the Fleet Air Arm and Australian naval flying before that and examines the careers of the Hawker Sea Fury, Fairey Firefly and de Havilland Sea Venom.

One of that select group of aircraft which might fall into the 'ultimate piston engined fighter' category, the Sea Fury was the Royal Australian Navy Fleet Air Arm's first carrier borne fighter, sharing the early operational responsibilities of the FAA with the Fairey Firefly.

Fast and manoeuvrable, the Sea Fury entered service at a time when the jet fighter was coming into its own, but nevertheless provided Australia with a useful naval capability which was put to the test in combat in the Korean war, operating from HMAS *Sydney*.

The Sea Fury remained the RAN's fighter until replaced by the Sea Venom in 1956 and in the meantime was even responsible for the shooting down of one aircraft, a runaway Auster which had taken off without its pilot!

The versatile Firefly was one of the truly great naval aircraft of the 1940 and 1950s, entering service with the Royal Navy towards the end of World War II and remaining in production for some years afterwards.

Developed into fighter, strike, anti submarine patrol, trainer and other versions, the Firefly comprised – along with the Sea Fury – half the RAN FAA's initial front line equipment and like its fighter counterpart flew operationally in Korea from the carrier HMAS *Sydney*.

One of the Firefly's main roles in Australian service was that of anti submarine patrol, but even after it had been replaced in that role by the Gannet it flew on for some years as a trainer and target tug.

Replacing the Sea Fury as the Royal Australian Navy's shipborne fighter, the Sea Venom was one of the series of distinctive twin tail boom de Havilland jet fighters which had begun with the Vampire in the 1940s and been developed through single and two seat Vampires and Venoms.

When it entered service with the RAN in 1956, the Sea Venom became Australia's first all weather jet fighter, pre-empting the RAAF by eight years before it got a fighter with that capability, the Mirage.

Flying from HMAS *Melbourne*, the Sea Venom served with the RAN for just over a decade before being replaced by the Douglas A-4 Skyhawk. In the meantime, the aircraft built up a solid reputation as a reliable workhorse.

Hopefully, what follows will provide some insight into the Australian use of these three important naval aircraft.

AUSTRALIAN NAVAL FLYING

Although the Royal Australian Navy did not gain government approval for the establishment of its own Fleet Air Arm until 1947 (along with the associated procurement of aircraft carriers), naval flying in Australia has a long history which precedes these events.

The sloop HMAS *Una* had become the first RAN ship to carry aircraft in November 1914 when a BE.2A landplane and Maurice Farman seaplane were transported to Rabaul but were not flown there, remaining in their packing cases until returned to Australia in early 1915.

Operational flying from the cruiser HMAS *Pioneer* using Short and Sopwith seaplanes for spotting duties in the search for the German warship *Konigsberg* were carried out off East Africa during 1915, these activities leading to requests for seaplanes to be embarked on other Australian cruisers.

These requests were denied as was one for the RAN to take over a British seaplane carrier.

Early Activities

Some progress was made in 1917 when the Australian cruiser HMAS *Brisbane* was operating from Colombo in company with the British seaplane carrier HMS *Raven II*. *Raven's* aircraft complement comprised four Shorts 184s and a Sopwith Baby. The Baby was transferred to *Brisbane* and the aircraft used in searches for the German raider *Wolf*. Its operation was successful, but again requests for permanent deployment of such aircraft aboard the RAN's cruisers were refused by the British Admiralty.

But not all was lost, an action by the famous cruiser HMAS *Sydney* (which in 1914 had sunk the German commerce raider *Emden*) in May 1917 underlined the need for shipboard aircraft. On this occasion, the German Zeppelin L43 had attempted to bomb the ship from a height to which its guns could not reach. *Sydney's* skipper, Captain J S Dumaresq, considered an aircraft could have intercepted the Zeppelin and therefore substantially reduced the risk to his own ship and to others which might be attacked later.

Dumaresq maintained his attitude that the carrying of aircraft by the cruisers should be regarded as an essential part of its armoury and had a revolving aircraft launching platform fitted to *Sydney* during a refit. Dumaresq's perseverance paid off, eventually. He was asked to supervise flying trials from the battlecruiser HMS *Repulse* and their success led to a change of thinking within the Admiralty. It was discovered that heavy cruisers and larger ships could have a platform fitted to the guns of one turret but that a separate revolving platform (such as that fitted to *Sydney*) would be needed for light cruisers.

HMAS *Sydney* became the first Royal Australian Navy ship to launch an aircraft from such a platform on 8 December 1917 when a Sopwith Pup was sent aloft. One further successful launch was carried out, and the scene was finally set for Australia's

Trial launching of a Sopwith 1½ Strutter from a gun platform on the battlecruiser HMAS Australia *on the Firth of Forth during 1918. (RAN)*

Australian naval flying received a boost in 1920 with the placing of an order for six Fairey IIID seaplanes. These and other types which followed officially belonged to the RAAF. This changed in 1948 with the establishment of the Fleet Air Arm.

The seaplane carrier HMAS Albatross *passes under the almost complete Sydney Harbour bridge, circa 1931.* Albatross *represented the first step towards a true Australian Naval Air Service.*

Supermarine Seagull IIIs were flown by the RAAF's No 101 Fleet Co-operation Flight which operated under the operational control of the Navy.

cruisers to have aircraft operating from them.

Further successful trials were carried out from HMAS *Australia* during the same month when a Pup was flown from the quarterdeck, the first time any aircraft had taken off from a ship's deck. Later trials saw flights from platforms built over the guns, and the result was that from early 1918 HMAShips *Australia*, *Sydney* and *Melbourne* each carried a Sopwith Camel for fighter duties and in the case of *Australia*, a Sopwith 1½ Strutter for reconnaissance missions as well.

The first action recorded by these aircraft was in June 1918 when Sydney and *Melbourne* were operating as part of a British cruiser squadron which raided Heligoland Bight in northern Germany. The raid was met by two German aircraft which tried to bomb the ships and the Australian cruisers' Camels were launched in an attempt to intercept them.

Flt Lt L B Gibson, in*Melbourne's* aircraft, lost sight of the enemy and returned, but Flt Lt A C Sharwood from HMAS *Sydney* pursued the aircraft and attacked one of them. The German was seen to be in a spinning dive, but the combination of jammed guns and other enemy aircraft forced him to return to the fleet, surviving 'friendly' fire from the squadron's ships before landing in the sea and being picked up.

Between the Wars

Even before World War I had ended, the value of aircraft to the Royal Australian Navy had been recognised and in June 1918 the Australian Naval Board recommended the formation of an Australian Naval Air Service and requested the loan of a seaplane carrier from Britain. This was once again refused (ostensibly because all such ships were fully occupied) and although Avro 504s were flown off Australian cruisers briefly, the combination of postwar cuts and infighting between the services over the establishment of what would become the Royal Australian Air Force slowed things down considerably and finally killed any hopes of a separate Naval Air Service.

There was some progress in Australian naval flying, however, beginning in 1920 with the placing of an order for six Fairey IIID seaplanes which although on the strength of the RAAF (as would all such aircraft until 1948) would be flown from the RAN's light cruisers for survey duties. The survey ship HMAS *Geranium* was also converted to carry a Fairey IIID, and overall, the aircrafts' operations were successful and laid the foundations for the future.

In June 1925, it was announced that the RAN would at last get its own seaplane carrier. HMAS *Albatross* would be built at Sydney's Cockatoo Island Dockyard and was originally intended to operate Fairey IIIDs. With the IIID's planned phasing out of RAAF service in 1928, these plans were revised and six (later nine) Supermarine Seagull III amphibians ordered instead.

These were given the RAAF serials A9-1 to A9-9 and the first examples were delivered in April 1926, the same month *Albatross* was laid down. In service, the Seagulls would be flown by the RAAF's No 101 Fleet Co-operation Flight which operated under the operational control of the Navy.

The seaplane carrier was launched in February 1928 and commissioned in January 1929. She was 443ft 9in (135.26m) long overall, displaced 4,877 tonnes standard and 6,096 tonnes deep load and carried up to nine aircraft and a crew of 450. Her pair of Parsons geared turbines produced 12,000 shaft horsepower between them and pushed her to a maximum speed of 22 knots. Four 4.7in (11.9cm) guns (along with 2-pounder anti aircraft and other lighter weapons) were fitted.

HMAS *Albatross* served with the RAN until 1938 when she was sold to the Royal

*The Supermarine Seagull V (or Walrus) amphibian was originally designed specifically to operate from HMAS*Albatross*. The type also flew from the RAN's cruisers and 61 were delivered between 1935 and 1944.*

The RAN operated three Supermarine Sea Otter amphibians between 1948 and 1953.

Navy for £266,500 ($523,000). Due to the effects of the Great Depression of the 1930s and the defence spending cuts which inevitably followed, *Albatross* spent some of the last five years of her RAN career in reserve.

One operational problem which had surfaced was the realisation that the Seagull III was basically unsuitable for the roles it was being asked to perform due to a lack of performance. As a result, Supermarine designed and built the Seagull V (or Walrus as it was known in British service) specifically for operations from *Albatross*.

There is therefore more than a little irony in the fact that when the first of 24 Seagull Vs (with RAAF serials A2-1 to -24) arrived in 1935 it was discovered the aircraft was too tall to fit into the ship's hangar, resulting in the need to build special trolleys on which the aircraft were placed with their undercarriages retracted! The Seagulls also served aboard the RAN's cruisers and *Albatross* was recommissioned in 1935 with the arrival of the new aircraft.

The RAN Fleet Air Arm's first front line equipment – Hawker Furies, Fairey Fireflies and the carrier HMAS Sydney. *The ship was commissioned in February 1949. (RAN)*

A further 37 Walruses were delivered for service from RAN cruisers between 1939 and 1944, these aircraft retaining their RAF serials.

As for *Albatross*, she retained that name in Royal Navy service, operating in the trade protection role off West Africa and in the Indian ocean. She was hit by a torpedo in 1944 and although not seriously damaged was never repaired. Postwar, she was converted into a merchantman and was scrapped at Hong Kong in 1964.

Towards A Carrier

Australian naval flying during World War II comprised mainly operations of the Walruses from the ships, and of course the contribution made by pilots from Australia and other Commonwealth nations to Britain's Fleet Air Arm.

The Pacific war proved the value of the aircraft carrier beyond doubt and by 1942 there was serious talk of obtaining such a ship for the Royal Australian Navy.

The ship in question was the elderly (commissioned in 1923) HMS *Hermes* of 11,000 tonnes, but even as her transfer to the RAN was being considered she was sunk by aircraft from the Japanese carriers *Soryu*, *Hiryu* and *Akagi* off Ceylon in April 1942.

The demand for aircraft carriers around the world meant the question would not be again addressed for another two years when plans were made to acquire the 13,400 tonne *Colossus* class carrier HMS *Ocean* from Britain, at the time four months away from launching in July 1944.

Although there was general agreement that Australia should acquire an aircraft carrier, there was vacillation as to which one, *Ocean* giving way to the nearly complete HMS *Venerable* (due to be commissioned in late 1944) and then returning to favour in mid 1945.

In the meantime there was much unresolved discussion about manning, paying for and establishing the logistics required for the ship. As a result, decisions were deferred and the project regarded as no longer being a 'war programme' but one to be considered in the light of postwar planning. For the moment, that was it, but it wasn't long after the cessation of hostilities that the issue was raised once again.

At Last, the Fleet Air Arm

The case for an Australian aircraft carrier and Fleet Air Arm was pushed hard in the last months of the war and immediately afterwards by Navy hierarchy. In September 1946 the British Admiralty offered Australia a deal which seemed too good to be true ... two of the still incomplete *Majestic* class carriers at a cost of £3,437,500 ($6.875m), or two for the price of one.

It was considered that the carriers should be equipped with the same aircraft types as the Royal Navy (the Sea Fury and Firefly) and that to sustain the package (two carriers and three Carrier Air Groups), a major Naval Air Station would have to be established. Training would be provided by Britain.

Despite the apparent generosity of this offer, it wasn't until 2 June 1947 that the Australian government authorised the Navy to implement a plan for naval aviation. Even now, it was undecided which service (RAAF or RAN) would provide the air component for this Naval Air Branch, but on 3 July 1947

A Royal Navy Sea Fury takes off from HMAS Sydney *during the carrier's working up trials.*

HMAS SYDNEY
***History**: Laid down (as HMS Terrible) 19 April 1943; launched 30 September 1944; ordered by RAN June 1947; commissioned (as HMAS Sydney) 5 February 1949; withdrawn from service 1973.*
***Builder**: Devonport Dockyard, South Devon, England.*
***Displacement**: 14,744 tonnes standard; 19,863 tonnes deep load.*
***Dimensions**: Overall length 698ft (212.75m); waterline length 650ft (198.12m); flight deck length 685ft (208.79m); beam (hull) 80ft (24.38m); beam (flight deck) 75ft (22.86m); draught (deep load) 24ft 6in (7.47m).*
***Machinery**: Two Parsons single reduction geared turbines; four Admiralty 3-drum boilers; 42,000 shaft horsepower.*
***Performance**: Max speed 24.5 knots (45.4km/h); service speed 20.5 knots (38km/h); endurance 9,320nm (17,260km) at 12 knots, 7,350nm (13,610km) at 20 knots (37km/h).*
***Bunkerage**: Fuel oil 3,230 tonnes; Avgas 80,000gal.*
***Armament**: Six twin and 16 single 40mm Bofors guns, reduced on conversion to fast transport.*
***Accommodation**: (Korea) 161 officers and 1343 men including aircrew; up to 37 aircraft comprising two squadrons of Sea Furies and one of Fireflies. Four Wessex helicopters during Vietnam deployments.*

HMAS Sydney *– with Fireflies and Sea Furies on deck – photographed in February 1955. (via Mike Kerr)*

the Navy's plan to operate its own aircraft was approved and the establishment of the Fleet Air Arm was finally assured.

The Ships

The carriers selected for the RAN were two of the six *Majestic* class vessels launched in 1944-45 but remaining incomplete after work on them was suspended in 1946. They were HMS *Terrible* which was commissioned into RAN service on 5 February 1949 as HMAS *Sydney* and HMS *Majestic*, which became HMAS *Melbourne* when she was commissioned on 28 October 1955.

When it was revealed *Melbourne's* commissioning would be nearly two years later than planned, it was decided to modify her to operate jet aircraft, further delaying completion but at least bringing her up to modern standards. The substantial modifications included the incorporation of a 5.5 degree angled flight deck, new arrestor gear, a steam catapult, larger deck lifts and a mirror landing system. Her first air group comprised de Havilland Sea Venoms and Fairey Gannets.

To help fill the void while *Melbourne* was being completed, the light fleet carrier HMS *Vengeance* was loaned to the Royal Australian Navy between November 1952 and October 1955.

HMAS *Sydney* remained straight decked throughout her career, arriving in Australia in May 1949 with the first batches of Sea Furies and Fireflies aboard. As will be described in a later chapter, these aircraft flew operationally from *Sydney* in the Korean conflict between September 1951 and January 1952.

Sydney was relegated to training duties in 1955 after government policy decided a single carrier was sufficient for Australia's needs (besides, she was not able to operate jets) and served in that role until 1958 when she was put into 'special reserve' –

Vengeance *and* Sydney *line astern, both with Sea Furies and Fireflies on deck.*

mothballed, but capable of being returned to service.

This occurred four years later following a 1960 decision when tensions associated with the period of 'confrontation' with Indonesia began, these in turn resulting from the formation of the Federation of Malaysia, a coalition of the former British colonies Malaya, Singapore (which later withdrew), Sarawak (North Borneo) and Sabah.

Sydney was therefore recommissioned in 1962 as a fast troop transport, combining this role with training and making numerous voyages to Vietnam in support of Australia's considerable involvement in that conflict. The aircraft carrier was finally decommissioned in 1972 and sold for scrap three years later.

HMAS *Melbourne* survived various threats to the future of the Fleet Air Arm until she was finally paid off in June 1982, fixed wing flying (with the A-4 Skyhawks and S-2 Trackers which had replaced the Sea Venoms and Gannets in 1967) ending two months earlier.

The conservative Liberal government of the day had intended replacing her with a ship and had even announced the purchase of the 'Harrier carrier' HMS *Invincible* until her sale was withdrawn following the Falklands war. But a change of government to the left wing Labor (*sic*) Party in March 1983 ended any thoughts of the Fleet Air Arm continuing to operate fixed wing aircraft from ships.

In fact the decision not to replace *Melbourne* was announced on the *very day* the new government was sworn in!

Two main shore establishments have been occupied by the Fleet Air Arm: HMAS *Albatross* at Nowra on the south coast of New South Wales (along with a satellite airfield at nearby Jervis Bay) and HMAS *Nirimba* at Schofields aerodrome west of Sydney. *Albatross* was and is the FAA's primary shore establishment and centre of flying from dry land, while *Nirimba* was also used for flying operations before becoming a general RAN training base.

The Royal Australian Navy Fleet Air Arm survives, but as a mainly helicopter equipped force operating from smaller ships in the anti submarine role with 16 Sikorsky Seahawk helicopters as its primary equipment.

(right) HMAS Vengeance *displays her code letter 'Q'. This photograph was taken in 1954. (L Kenderdine)*

The RAN's first helicopter was the Bristol Sycamore. 13 of which were delivered from 1953.

The Fleet Air Arm's second front line duo was the Sea Venom all weather fighter and Fairey Gannet anti submarine aircraft. They are photographed aboard HMAS Melbourne. *(via Steve Long)*

(left) HMAS Melbourne *with the third generation of front line aircraft aboard – The Douglas Skyhawk and Grumman Tracker, with a quartet of Westland Wessexes. (RAN)*

HMAS MELBOURNE
***History**: Laid down (as HMS Majestic) 15 April 1943; launched 28 February 1945; ordered by RAN June 1947; commissioned (as HMAS Melbourne) 28 October 1955; paid off 30 June 1982.*
***Builder**: Vickers-Armstrong, Barrow-in-Furness, Cumbria, England.*
***Displacement**: 16,256 tonnes standard; 20,645 tonnes deep load.*
***Dimensions**: Overall length 701ft 6in (213.82m); waterline length 650ft (198.12m); flight deck length 690ft (210.31m); beam (hull) 80ft (24.38m); beam (flight deck) 105ft (32.0m); draught (deep load (25ft (7.72m).*
***Machinery**: Two Parsons single reduction geared turbines; four Admiralty 3-drum boilers; 42,000 shaft horsepower.*
***Performance**: Max speed 24.5 knots (45.4km/h); endurance 12,000nm (22,200km) at 14 knots (26km/h).*
***Bunkerage**: Fuel oil 2,948 tonnes; Diesel 356 tonnes; Avtur 140,171gal; Avgas 3,744gal.*
***Armament**: (early) six twin and 13 single 40mm Bofors guns; (late) four twin and four single 40mm Bofors guns.*
***Accommodation**: Crew of 1,210; aircraft 22-27. Sea Venom and Gannet (1956), Wessex (1963), A-4 Skyhawk and S-2 Tracker (1967), Sea King (1975).*

The Fairey Gannet anti submarine aircraft, 36 of which were delivered 1956-58.

A Skyhawk prepares for launch from HMAS Melbourne *while a Tracker waits in the background. (RAN)*

The Fleet Air Arm's current front line aircraft, the Sikorsky Seahawk anti submarine helicopter.

HAWKER
SEA FURY

This low ground shot of Royal Australian Navy Sea Fury WZ646 epitomises the ruggedly powerful nature of what many regard as being the 'ultimate' piston engined fighter.

HAWKER SEA FURY

Belonging to that select group of late or postwar piston engined fighters often referred to as the 'ultimate' of their *genre* the Sea Fury was typical of many of the aircraft to emerge from the Hawker stable.

Evolutionary rather than revolutionary, its origins can be traced back to the Typhoon which first flew in February 1940 via the Tempest (first flight September 1942), itself developed as a faster 'thin wing' variant of the Typhoon.

The Fury evolved as a 'Light Tempest' and the substantially redesigned result was an aircraft offering very high performance and representing the pinnacle of piston engined fighter design. Developed for the Royal Air Force, the first Fury flew in September 1944, a time when the world's leading air forces were already preparing for the introduction of jet fighters.

As a result, by far the most numerous Fury variants were navalised Sea Furies, the Royal Navy's last piston engined fighter and adopted by that service as a result of it not yet being fully committed to the jet engine.

The first navalised Fury was flown in February 1945 and 860 Furies and Sea Furies of all versions were built of which 615 were delivered to the Royal Navy. Export customers were Australia, Iraq, Egypt, Pakistan, Canada, The Netherlands (some built under licence by Fokker), West Germany, Burma and Cuba.

The Royal Australian Navy took delivery of 101 Sea Furies between 1949 and 1953 and operated them in combat during the Korean War. The aircraft served as the RAN Fleet Air Arm's primary shipboard fighter until replaced by the Sea Venom in 1956. The last Sea Fury was struck off RAN charge in 1963, although the vast majority of the survivors had been disposed of by 1960.

The Hawker Heritage

The H G Hawker Engineering Company Ltd was established in November 1920 largely to replace Sopwith Aviation, the company which had provided some of the Great War's most notable fighter aircraft but had folded earlier in 1920 after the British Treasury had served an enormous claim for Excess War Profits Duty.

Hawker took over Sopwith's premises at Canbury Park Road, Kingston-upon-Thames, in the south-west part of the Greater London area.

The company was named after the Australian pilot Harry Hawker, who had joined Sopwith in 1912 as a test pilot and who subsequently achieved fame as a pioneer. Hawker was appointed one of the original directors of the company which bore his name, but his life ended tragically in July 1921 when the Nieuport Goshawk he was testing for a race crashed, the result of Hawker suffering a haemorrhage while pulling heavy 'g' loadings in tight turns. Hawker was 32 years old.

But the company carried on, its golden years occurring during the 1925 to 1959 period when Sir Sydney Camm was the chief designer. The first new design to be launched under Camm's leadership was the Hornbill light fighter of 1926, the ancestor of the family of single engined biplane day bombers, fighter/bombers, army co-operation and general purpose military aircraft beginning with the Hart of 1928 and progressing through the 1930s with the Nimrod, Demon, Osprey, Audux, Hardy, Hind and other variants.

Following the same basic design was the smaller, single seat Fury fighter of 1931. Like its bomber counterparts, the Fury was ordered in quantity for the Royal Air Force and foreign customers.

Hawker became a public company in 1933 and assumed the new title Hawker Aircraft Ltd. On its flotation, the new company took over the assets of the old, which was eventually dissolved.

In 1934, Hawker took over another famous British aviation name, Gloster Aircraft, although the two companies retained separate identities until 1963 when they and other companies – including de Havilland, Armstrong Whitworth, Blackburn, Avro and Folland – were grouped together under the Hawker Siddeley banner.

In November 1935 Hawker flew the prototype of possibly its best known product, the Hurricane. The RAF's first monoplane fighter was typically evolutionary in its design, utilising the structural principals of the Hart/Hind/Fury series – all metal and fabric covered, the fuselage being of steel tube

Sea Fury FB.11 WH587 of the RAN, complete with underwing ordnance shackles. This Sea Fury remains airworthy in the USA.

An early production Sea Fury photographed over the Portsmouth area in the late 1940s. (BAe)

design – married to a Rolls-Royce Merlin engine and retractable undercarriage.

The RAF's first 300mph (483km/h) and eight gun fighter, the Hurricane soon acquired metal stressed skin wings and was developed into fighter/bomber, naval fighter and specialist ground attack variants. Its finest hour was, of course, the Battle of Britain in 1940, where it and smaller numbers of Spitfires helped saved the day.

The Hurricane remained in production until 1944 when the 14,527th example was delivered. Of those, 1,077 were built by the Canadian Car and Foundry Company.

The Hurricane, one of a long line of classic Hawker fighters. This particular example (a Mk.XII) belonged to the Canadian Warbird Heritage but was destroyed in a hangar fire at Hamilton, Ontario in February 1993. (Jim Thorn)

Typhoon and Tempest

The road to the Sea Fury began in 1937 when Sydney Camm and his team were investigating the development of a new 12 gun fighter powered by the new Napier Sabre 24 cylinder engine of horizontal 'H' configuration which promised power outputs of more than 2,000 horsepower.

The British Air Ministry issued Specification F.18/37 to cover such an aircraft, or specifically two, one powered by the Sabre (Type N) and the other by the Rolls-Royce Vulture 24 cylinder 'X' type engine (Type R). These fighters were named Typhoon and Tornado, respectively, and were similar designs apart from their powerplants and other details. Structurally, they offered an advance over the Hurricane by featuring a stressed skin semi monocoque rear fuselage design in combination with the traditional steel tube forward fuselage.

Problems with the Vulture and the pressures of Hurricane production in 1940 contributed to the Tornado being cancelled, efforts then being concentrated on the Typhoon.

This aircraft also suffered numerous teething problems including unreliability from the 2,100hp Sabre I engine originally fitted, vibration, poor low speed handling and several failures of the monocoque structure behind the cockpit.

Three RAAF Hawker Demons in formation. The Demon was one of a prolific family of Hawker light bombers, fighter/bombers, trainers and army co-operation aircraft of the late 1920s and 1930s which included the Hart, Hind, Audax and others.

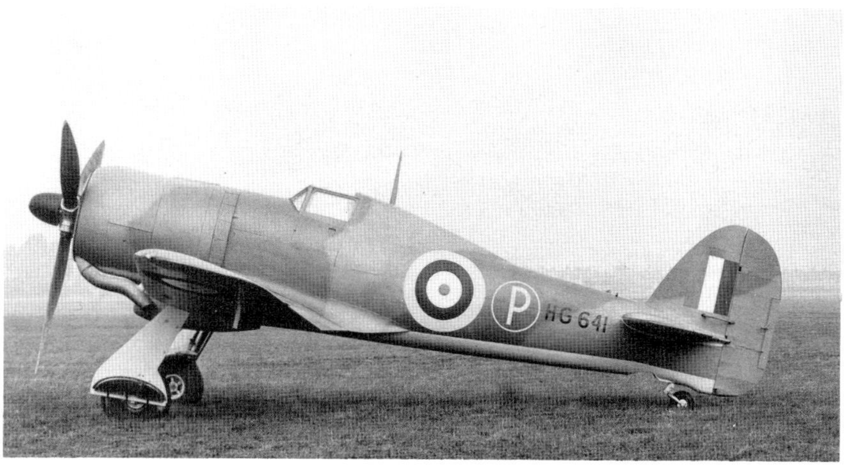

Despite these and a six month delay during 1940 while the production of Hurricanes had priority, the Typhoon entered service in 1941 with improved engines but lingering problems with the structural integrity of its rear fuselage, some 135 failures of which were reported before the problem was solved. The Sabre engine also continued to give some problems, and for a time many more Typhoons were lost due to these causes than for any other reason including enemy action.

Additionally, the Typhoon's climb and altitude performance were poor, making it unsuitable as a fighter, and the aircraft didn't find its *niche* until after D-Day when, fitted with rocket projectiles or bombs, it became an effective ground attack fighter, roaming across France and other parts of Europe on offensive sweeps against armour, trains and other targets on the ground.

The Typhoon's thick wing made it susceptible to the effects of compressibility at higher speeds, so by 1941 Sydney Camm had begun work on 'thin wing' development of the fighter.

Originally known as the Typhoon II, the new aircraft featured numerous changes

(left) The Hawker Tornado (top) and Typhoon, built to a similar specification but with different engines. The Typhoon went into production but the Tornado's original powerplant, the Rolls-Royce Vulture, was unsuccessful. The Tornado illustrated here has been re-engined with a Bristol Centaurus radial engine as used in the later Tempest II and Sea Fury. (BAe)

over its predecessor and was renamed Tempest. The Tempest's thinner, semi elliptical wing meant moving some fuel capacity to the fuselage, resulting in a lengthening of that part of the aeroplane. Structurally, the Tempest was similar to the Typhoon.

Tempests were developed to be powered by three different engines, the Napier Sabre, Bristol Centaurus 18 cylinder radial and Rolls-Royce Griffon V12.

Production centred around the wartime Tempest V and the too late to see WW2 service Tempest VI (both with Sabre engines) plus the Centaurus powered Tempest II, which also entered service too late to see action during the war. The Tempest II was generally regarded as being the best of the bunch and it was on this aircraft that early thoughts on a Centaurus powered 'Tempest Light Fighter' were based.

Of the 1,399 Tempests manufactured, 805 were Mark Vs. The type entered service in April 1944 and soon proved itself to be the fastest fighter low down, topping 392mph (631km/h) at sea level and a remarkable 416mph (669km/h) at 4,600 feet.

Although no slouch as a fighter or ground attack aircraft, the Tempest really made its name defending Britain against the V-1 'flying bombs' being launched against it from Europe. During the June to September 1944 period, no fewer than 638 V-1s were accounted for by Tempests.

Immediate predecessor of the Fury and Sea Fury was the Tempest, illustrated here by the one-off Tempest I with Napier Sabre IV engine (top) and the Tempest V (bottom) with the more reliable Sabre II. The latter was the major production version of the Tempest, but the Centaurus powered Tempest II was also built in quantity. (BAe)

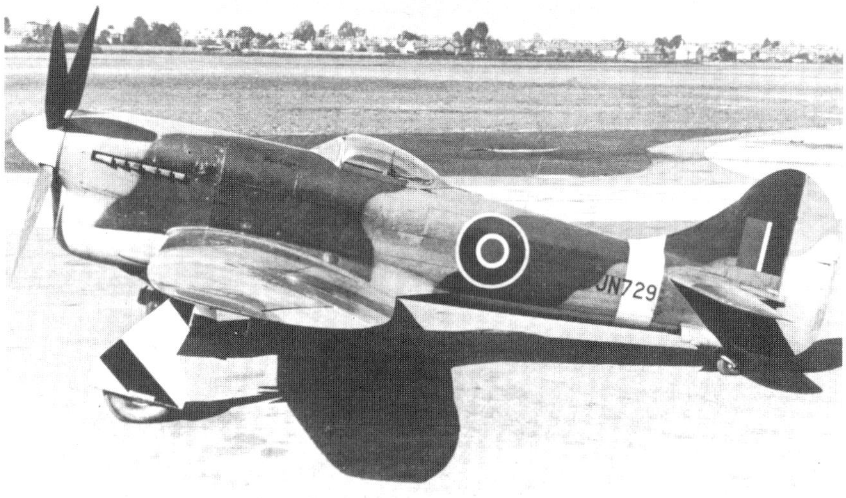

A Lighter Fighter

A combination of Sydney Camm's efforts to investigate a 'Tempest Light Fighter' with Bristol Centaurus radial engine and the accidental arrival in Britain of a *Luftwaffe* Focke-Wulf Fw 190 fighter were the major catalysts in the development of what would become the Hawker Fury and Sea Fury.

Even before the first Tempest had flown, Hawker's design team was investigating a new fighter intended to halt the trend towards ever larger and heavier fighters. These early investigations were begun in 1941 but remained design studies.

It wasn't until the second half of 1942 that firm ideas began to emerge, based around the elimination of the Tempest's wing centre section for a wing of reduced span and the use of a fully monocoque fuselage structure instead of the combined steel tube/monocoque structure previously employed.

In the meantime, an amazing stroke of good fortune had come Britain's way. On 23 June 1942, Oberleutnant Armin Faber somehow managed to mistake RAF Pembrey (south Wales) for a *Luftwaffe* Channel coast airfield and landed there, presenting the RAF with an intact Fw 190A-3. The 190's compact design, efficient radial engine installation and light weight had a profound influence on many, and was primarily responsible for the issue of Air Ministry specification F.6/42 which was tailor-made for Hawker's 'Light Tempest' studies.

The prototype Hawker F.2/43 Fury, NX798, first flew on 1 September 1944 powered by a Bristol Centaurus engine as would be fitted to all production Furies and Sea Furies.

LA610, the second prototype Fury which first flew in November 1944 powered by a Rolls-Royce Griffon inline engine. It was subsequently used to test Sabre (as here) and Centaurus powerplants (BAe)

A new specification – F.2/43 – was written around the project and issued in March 1943. Another specification entered the equation the following month when N.7/43 was issued for a naval interceptor. Camm and Hawker proposed the same design to meet both specifications with the naval fighter powered by an uprated Centaurus XII engine.

This proposal was accepted but in order to avoid overloading Hawker's already heavy production schedules, the Boulton Paul company was appointed the main contractor for the naval version. A new specification (N.22/43) was subsequently written for the production version of what would become the Sea Fury.

Orders and Into the Air

Later in 1943, orders were placed for six prototypes of Hawker's new fighter, three to be powered by Bristol Centaurus radial engines, two by Roll-Royce Griffon inline V12s and one as a structural test specimen. Only one of the Griffon-powered prototypes was actually built and it was subsequently re-engined with Centaurus and Napier Sabre powerplants.

The first production orders were placed in April 1944 and covered 200 land based and 200 naval fighters for the RAF and Royal Navy. Boulton Paul would be responsible for production of 100 of the naval aircraft. Later in the year, the aircraft were formally named Fury Mk.I and Sea Fury Mk.X, respectively.

The project was immediately facing problems, however, as the entire Royal Air Force order was cancelled in January 1945 thanks to the influence of the jet fighter and at the same time the Royal Navy contract was halved to 100 aircraft, eliminating the need for Boulton Paul to be involved.

Testing of the prototype Furies had meanwhile begun, the first aircraft, NX798, flying from Hawker's facility at Langley in the hands of test pilot Phil Lucas on 1 September 1944. NX798 was powered by a 2,400hp rigidly mounted Centaurus XII driving a four bladed propeller, an installation which would cause problems with vibration at low engine speeds and propeller overspeeding with the sudden advancement of the throttle – such as during a missed approach. This combination, along with engine crankshaft failures, kept the test pilots busy with forced landings. There were substantial early problems with the Centaurus, most of them resulting from faulty lubrication systems.

The second Fury (LA610) was flown 27 November 1944, initially powered by a 1,915hp Rolls-Royce Griffon 85 with a Rotol six bladed contra-rotating propeller. The cancellation of the RAF order resulted in this aircraft's redundancy and it was subsequently used as an engine test bed,

The first navalised Sea Fury flew in February 1945. Although fitted with a tail hook it lacked folding wings.

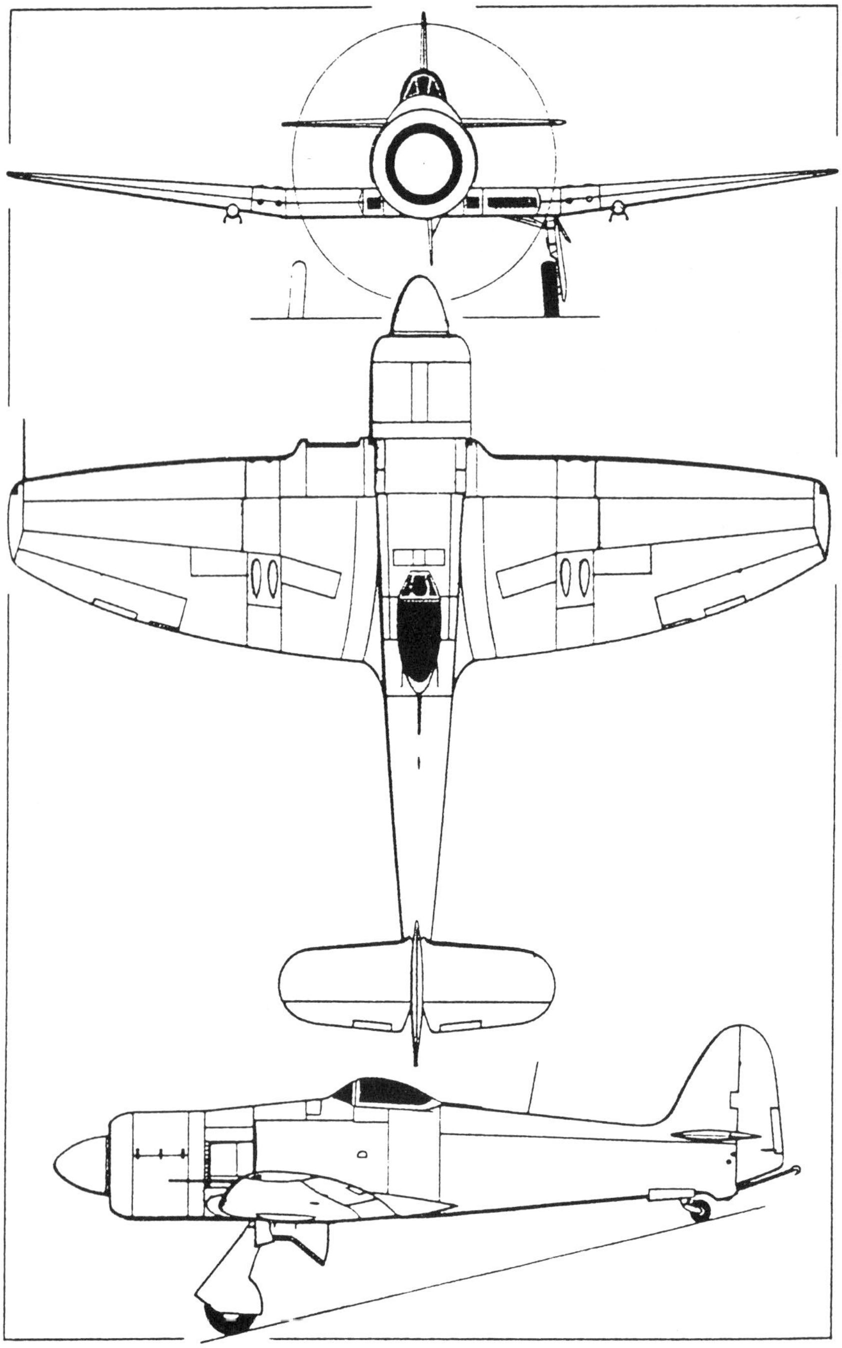

Hawker Sea Fury FB.11

HAWKER SEA FURY FB.11
Powerplant: *One Bristol Centaurus XVIII 18 cylinder two-row, sleeve valve, air cooled radial engine with single-stage two-speed supercharger rated at 2,480hp for t-o, 2,550hp at 4,000ft and 2,280hp at 16,500ft. Rotol five-bladed constant-speed propeller of 12ft 9in (3.89m) diameter. Internal fuel capacity of 200imp gal (909 l) in fuselage and wing tanks; provision for two 45imp gal (204 l) or 90imp gal (409 l) drop tanks.*
Dimensions: *Wing span 38ft 4.75in (11.69m); length 34ft 8in (10.57m); height 15ft 10.5in (4.84m); wing area 280sq ft (26.01sq m).*
Weights: *Empty 9,240lb (4,191kg); normal loaded 12,350lb (5,602kg); max t-o 14,650lb (6,645kg).*
Armament: *Four 20mm Hispano Mk.V cannon in wings with 145rpg; underwing armament options: two 500lb (227kg) or 1,000lb (454kg) bombs; twelve 3in (7.62cm) or 5in (12.7cm) 60lb (27kg) rockets; four 180lb (82kg) Triplex rockets.*
Performance: *(at 12,350lb/5,602kg) Max speed 460mph (740km/h) at 18,000ft, 415mph (668km/h) at 30,000ft; initial climb 4,320ft (1,316m)/min; time to 30,000ft 10.8min; service ceiling 35,800ft (10,911m); range (internal fuel) 700 miles (1,126km), range (with 90gal drop tanks 1,040 miles (1,674km).*

initially fitted with a Centaurus and then with a 3,055hp Napier Sabre VII which drove it to the very respectable maximum speed of 485mph (780km/h) at 16,500 feet.

The third Fury prototype, NX802 with Centaurus XV, was flown in June 1945, five months after the RAF order had been cancelled and subsequent to the first Sea Fury taking to the air. This aircraft was repurchased by Hawker shortly afterwards and used for development flying before being sold to Pakistan in 1949.

A fourth prototype (VP207) with Sabre VII engine was built in 1946 with the hope of encouraging export orders but the unreliable Sabre was always going to inhibit such activity and development of this branch of the Fury family soon ended. All production Furies and Sea Furies would be powered by the Bristol Centaurus.

Sea Fury Development

The first Sea Fury was SR661, powered by the same rigidly mounted Centaurus XII/Rotol four bladed propeller combination as NX798, was first flown on 21 February 1945 and completed to almost full navalised standard, lacking only folding wings.

The vibration and lubrication system problems first exhibited by NX798 were carried through to this first Sea Fury, and testing at Farnborough in May 1945 of the aircraft's rudder control for carrier operations showed up strong swinging tendencies on takeoff and directional instability during the landing roll. The result was the eventual fitting of a taller fin and rudder of greater area and the recommendation by test pilot Lt Cdr (later Captain) Eric 'Winkle' Brown that the fully castoring tailwheel should be fitted with a lock.

This was at first not accepted, but a subsequent incident in the second Sea Fury prototype convinced the powers-that-be that it was a good idea (see below).

SR661 did receive some other modifications including the fitting of a five bladed propeller and this, in combination with the larger vertical tail surfaces, made the aircraft rather more pleasant for carrier operations. After some arrested landing trials on dry land at Farnborough (which resulted in the fitting of a longer tail hook), SR661 flew initial carrier trials aboard HMAS *Ocean* during August 1945.

The second Sea Fury (SR666) was flown in the same month, powered by a Centaurus XV and fitted with a five bladed propeller from the start. It also featured the taller fin and rudder and was fully navalised with folding wings and the Sea Fury's standard fixed armament of four 20mm Hispano Mk V cannon, two in each wing.

SR666 participated in catapult testing at Farnborough in mid 1946, and it was during one of these tests that 'Winkle' Brown demonstrated the need for a lockable tailwheel.

The tail down method of launching was used, and on the very first launch the castoring tailwheel caused considerable drama. At takeoff, full left rudder was applied to counteract the aircraft's known tendency to strongly swing the other way. Regardless, the aircraft swung violently anyway, causing the castoring tailwheel to

A Sea Fury F.X from the first production batch of 50 aircraft. After that, production concentrated on the Sea Fury FB.11. (BAe)

Sea Fury FB.11 VW633 catches the wire on landing aboard an aircraft carrier. (RAN)

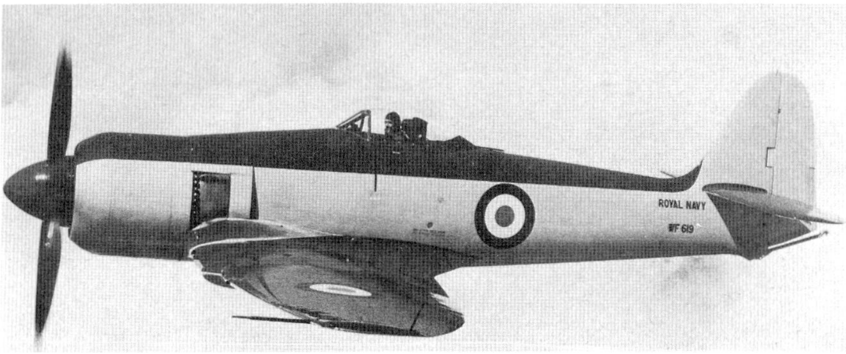

Various Sea Furies in flight and on the ground including one illustrating the relatively simple wing folding mechanism. (via Joe Barr)

hit a deck fitting. This stopped the swing, but the Sea Fury then began to yaw to port, all the while under the irresistible influence of the catapult.

The swinging rear fuselage and tailwheel narrowly missed some other obstacles, and aware of damage to his aircraft, Brown flew it to Hawker's Langley airfield and landed it on the grass, holding up the tail as long as possible. From then, Sea Furies had lockable tailwheels ...

A third Sea Fury prototype (VB857) was first flown in January 1946 and powered by a Centaurus XXII with an improved lubrication system and – most importantly – flexible dynafocal engine mountings which eliminated the vibration problems encountered with the original mounting design. These were incorporated in production aircraft.

Work on VB857 had been begun by Boulton Paul, but on the cancellation of that company's contract, the aircraft was completed by Hawker. By the time of its first flight, production Sea Furies were already in the early stages of manufacture at Hawker's Kingston facility.

Into Production

After what had been a reasonably lengthy gestation period, the first of 50 production Sea Fury F Mk.Xs – TF895 – flew in September 1946. In common with the nine aircraft which followed, this was fitted with a four bladed propeller initially but five bladers followed as soon as they were available. The other Sea Fury Xs were fitted with the more efficient propeller on their manufacture.

Most of the early aircraft were used for trials of some sort and the first RN Fleet Air Arm operational units to receive the aircraft were No 778 Squadron in February 1947 and 787 Squadron the following month. As both these squadrons were initially given training and development tasks, it is perhaps strange to note that the first front line unit to operate the Sea Fury (from August 1947) was No 803 Squadron, Royal Canadian Navy, based in Northern Ireland!

Despite the Canadians' lead, the first squadron to take the aircraft to sea was the Royal Navy's No 807, which embarked on HMS *Implacable* during May 1948. The Canadians got their sea legs a fortnight later, flying from HMCS *Magnificent*.

Armed only with four 20mm Hispano cannon along with 580 rounds of ammunition, the Sea Fury X was very much a specialist air superiority fighter, albeit with very high performance from its 2,550hp Centaurus XVIII engine. The aircraft could achieve a maximum speed of 465mph (748km/h) at 18,000 feet, climb at better than 4,000 feet per minute (1,200m/min) at normal loaded weight and fly some 700 miles (1,125km) on the internal fuel tankage of 200imp gal (909 litres).

Despite its similarities to the Centaurus powered Tempest II, the Sea Fury incorporated much redesign, not the least of which was the full monocoque fuselage design, under which was attached the two spar wing, similar to the Tempest's but without the centre section. This reduced the span

Two former Iraqi Hawker Furies which found their way to Australia in the early 1980s but are now overseas: Bruce Andrews' (top) and Ted Allen's aircraft now reside in the United Kingdom and the USA, respectively. (David Wilson/ Bruce Chapman)

Two views of Australia's only flying Fury in 1993, Guido Zuccoli's former Iraqi aircraft. (Bruce Chapman/Les Bushell)

by about 2ft 7in (78.7cm) to 38ft 4.75in (11.70m), while at 34ft 8in (10.56m), the Sea Fury was just three inches (7.6cm) longer overall than the Tempest II. As for the 'Lighter Tempest' concept, the Sea Fury's empty weight was actually some 300lb (136kg) greater than the Tempest II, largely due to the incorporation of the equipment necessary for operations from aircraft carriers.

Fuel was carried in a 97imp gal (441 l) main fuselage tank, a 30imp gal (136 l) auxiliary fuselage tank (both of these immediately forward of the cockpit), two 28imp gal (127 l) inter spar wing tanks and a 17imp gal (77 l) tank in the starboard wing leading edge. A 45 or 90imp gal (204 or 409 l) drop tank could be carried under each wing.

The structure was entirely metal and covered with metal skinning, as were all the control surfaces. The Tempest had a fabric covered rudder.

The typically Hawker wide track inward retracting undercarriage came from Dowty and was hydraulically operated, as were the wing folding mechanism and split trailing edge flaps in four sections.

The 18 cylinder Centaurus air cooled radial engine of 53.6 litres (3,270cu in) capacity was closely cowled and featured a single stage/two speed supercharger which produced 2,480hp for takeoff, 2,550hp at 4,000 feet and 2,280hp at 16,500 feet. The five bladed Rotol constant-speed propeller was of 12ft 9in (3.89m) diameter.

Model Most Numerous

The Sea Fury F.X remained in front line service for only a short time, most having been replaced by the more versatile FB Mk.11 by the end of 1948. Some F.Xs served with second line units for some time afterwards. With the new model came a change in the designation style from Roman to Arabic numerals.

The designation prefix 'FB' also indicated new roles for the Mk.11 as a fighter-bomber rather than straight fighter. On their withdrawal from service, many of the Mk.Xs were repurchased by Hawker as a source of spare parts.

The 51st production Sea Fury (TF956) was the first FB.11, and it was this variant which was produced in by far the largest numbers, some 565 of them being ordered by the Royal Navy and built between 1946 and 1954. The demands of the Korean War were responsible for the relatively large numbers of FB.11s, and 101 of them served with the Royal Australian Navy from 1949.

The major differences between the Sea Fury FB.11 and its predecessor were confined to the fitting of underwing hardpoints which enabled the aircraft to carry various items of ordnance. These included two 500lb (227kg) or 1,000lb (454kg) bombs, up to twelve 3in (7.62cm) or 5in (12.7cm) rockets with 60lb (27kg) warheads (on three double launchers per wing) or four triplex rockets with 180lb (82kg) warheads. In addition, two 45imp gal (204 l) or 90imp gal (409 l) drop tanks could be carried, the latter increasing the range to 1,040 miles (1,673km).

Two approaches to Sea Fury and Fury two seaters: the British T.20 (top) with separate canopies connected by a transparent 'tunnel' and those supplied to Iraq (centre) and Pakistan (bottom) with no connection between the two canopies.

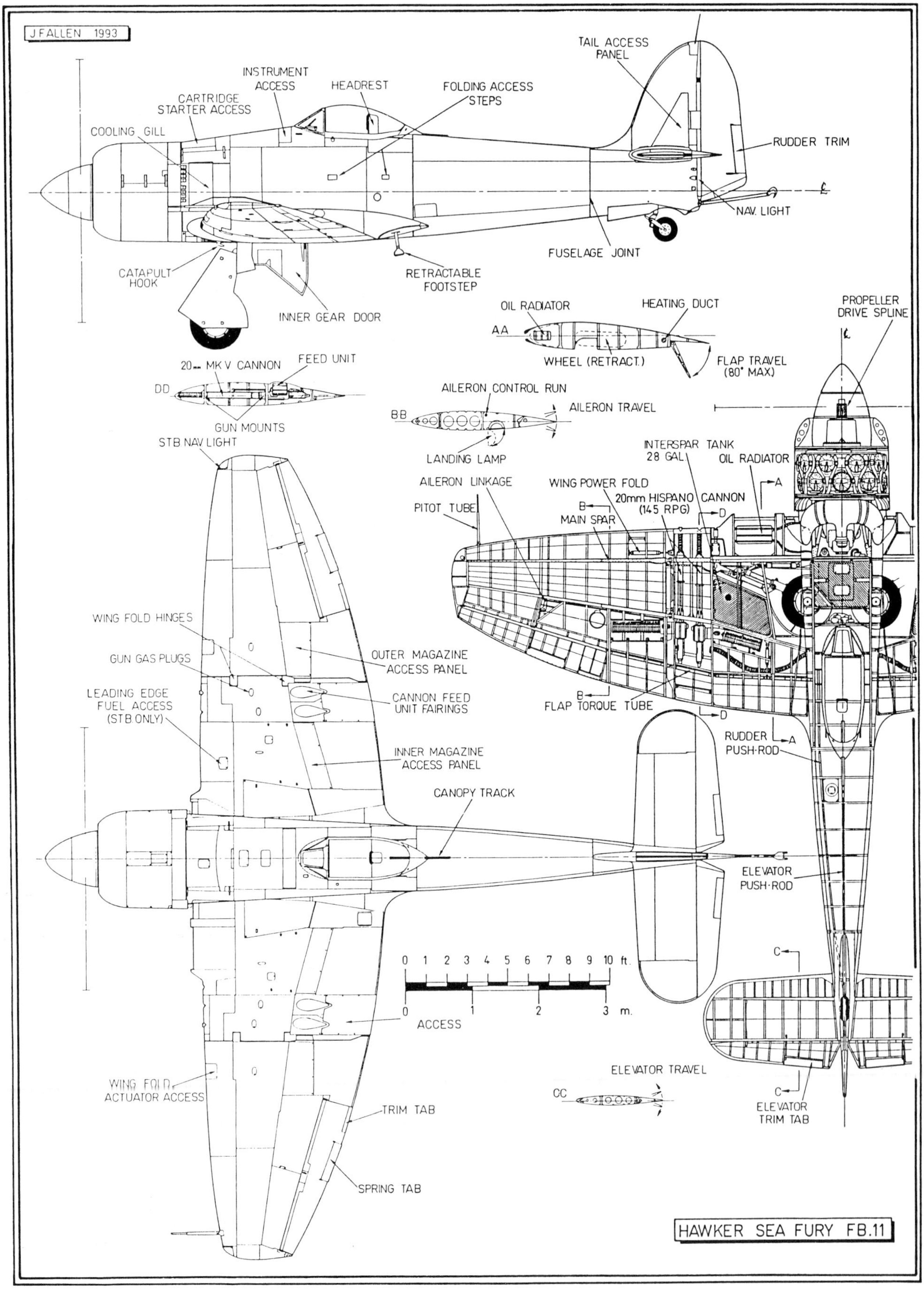
J FALLEN 1993
TAIL ACCESS PANEL
INSTRUMENT ACCESS
HEADREST
FOLDING ACCESS STEPS
CARTRIDGE STARTER ACCESS
COOLING GILL
RUDDER TRIM
NAV. LIGHT
FUSELAGE JOINT
CATAPULT HOOK
RETRACTABLE FOOTSTEP
INNER GEAR DOOR
OIL RADIATOR
HEATING DUCT
PROPELLER DRIVE SPLINE
AA
WHEEL (RETRACT.)
FLAP TRAVEL (80° MAX)
20mm MK V CANNON
FEED UNIT
DD
GUN MOUNTS
AILERON CONTROL RUN
BB
AILERON TRAVEL
LANDING LAMP
STB NAV LIGHT
INTERSPAR TANK 28 GAL
OIL RADIATOR
AILERON LINKAGE
WING POWER FOLD
20mm HISPANO CANNON (145 RPG)
PITOT TUBE
MAIN SPAR
WING FOLD HINGES
GUN GAS PLUGS
OUTER MAGAZINE ACCESS PANEL
LEADING EDGE FUEL ACCESS (STB ONLY)
CANNON FEED UNIT FAIRINGS
FLAP TORQUE TUBE
RUDDER PUSH-ROD
INNER MAGAZINE ACCESS PANEL
CANOPY TRACK
ELEVATOR PUSH-ROD
0 1 2 3 4 5 6 7 8 9 10 ft.
0 1 2 3 m.
ACCESS
ELEVATOR TRAVEL
CC
WING FOLD ACTUATOR ACCESS
ELEVATOR TRIM TAB
TRIM TAB
SPRING TAB
HAWKER SEA FURY FB.11

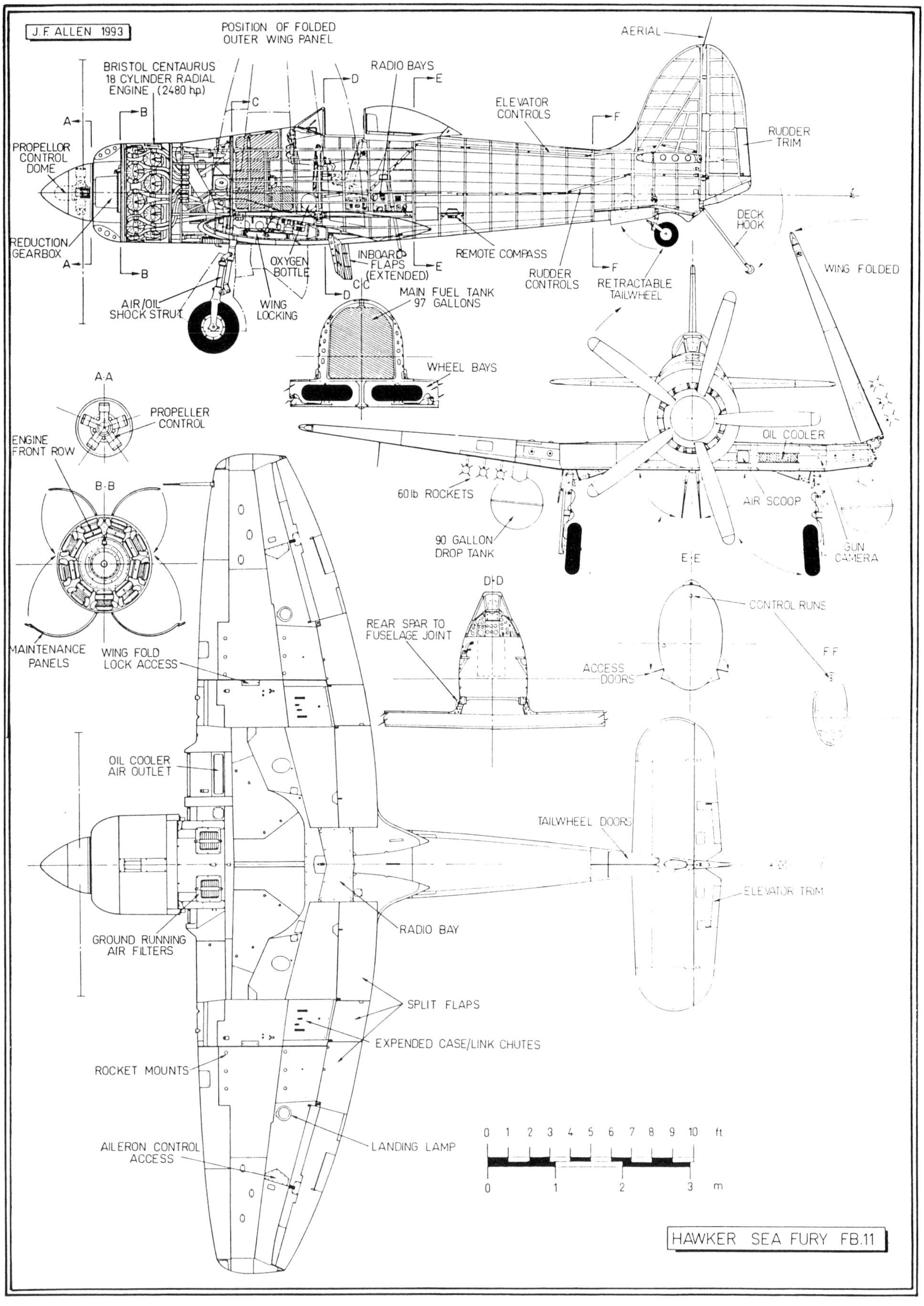
J.F. ALLEN 1993
POSITION OF FOLDED OUTER WING PANEL
AERIAL
BRISTOL CENTAURUS 18 CYLINDER RADIAL ENGINE (2480 hp)
RADIO BAYS
ELEVATOR CONTROLS
RUDDER TRIM
PROPELLOR CONTROL DOME
REDUCTION GEARBOX
DECK HOOK
REMOTE COMPASS
INBOARD FLAPS (EXTENDED)
OXYGEN BOTTLE
RUDDER CONTROLS
RETRACTABLE TAILWHEEL
WING FOLDED
AIR/OIL SHOCK STRUT
WING LOCKING
MAIN FUEL TANK 97 GALLONS
WHEEL BAYS
A-A
PROPELLER CONTROL
ENGINE FRONT ROW
B-B
OIL COOLER
60lb ROCKETS
AIR SCOOP
90 GALLON DROP TANK
GUN CAMERA
E-E
D-D
CONTROL RUNS
MAINTENANCE PANELS
WING FOLD LOCK ACCESS
REAR SPAR TO FUSELAGE JOINT
ACCESS DOORS
F-F
OIL COOLER AIR OUTLET
TAILWHEEL DOORS
ELEVATOR TRIM
GROUND RUNNING AIR FILTERS
RADIO BAY
SPLIT FLAPS
EXPENDED CASE/LINK CHUTES
ROCKET MOUNTS
LANDING LAMP
AILERON CONTROL ACCESS
0 1 2 3 4 5 6 7 8 9 10 ft
0 1 2 3 m
HAWKER SEA FURY FB.11

Sea Furies are popular in unlimited racing in the USA, but some have been converted to other powerplants. Frank Sander's T.20 'Dreadnought' racer is an example with a 28-cylinder Pratt & Whitney R-4360 Wasp Major replacing the original Centaurus. (Paul Merritt)

The FB.11 also had provision for rocket assisted takeoff gear (RATOG) and its normal loaded weight was increased from 10,660lb (4,835kg) to 12,350lb (5,602kg) and maximum weight to 14,650lb (6,645kg).

The decision to make the Sea Fury into a fighter-bomber came about as a result of the Royal Navy's need for such an aircraft, and the fighter was assessed as having considerable potential in that role. The later marks of the Supermarine Seafire were used to perform the RN's basic fighter role.

Trials with external stores had been carried out even before the first Mk.Xs had been delivered using SR666, the third Sea Fury prototype, and then TF923, a production Mk.X. The third Sea Fury prototype (VB857) was used for RATOG trials at Farnborough.

Sea Fury FB.11 deliveries began in May 1948, 802 Squadron being the first to receive the new model.

Sea Fury Versus MiG

The Sea Fury FB.11 was extensively used in the Korean War by Britain, Canada and Australia, flying off light fleet carriers and operating alongside Fairey Fireflies. (Australian Sea Fury and Firefly Korean operations are described in detail in a later chapter). The Sea Furies flew Combat Air patrol (CAP) and air-to-ground strike missions, often in the extremely difficult conditions presented by the Korean climate and against strong ground defences.

Additionally, the Sea Fury was a piston engined fighter which found itself sometimes pitted against MiG-15 jets, and the result was not often what might be expected due to the skills of the Sea Fury pilots and the ruggedness of the aircraft's design.

What was probably the Sea Fury's finest operational moment occurred on 9 August 1952 when four aircraft from the Royal Navy's No 802 Squadron (aboard HMS *Ocean*) took on eight MiG-15s and emerged not only intact but with one MiG destroyed and two others damaged to their credit. The pilot of the victorious Sea Fury was Lt P Carmichael, the flight leader.

On that day, the Sea Furies were flying an armed reconnaissance over North Korea when they were 'bounced' by the eight MiGs. A five minute dog fight ensued, the Sea Furies' break turns preventing the MiG pilots from getting good firing opportunities.

Several head on passes followed, the Sea Furies registering numerous hits. Then the MiGs withdrew and as they were departing an aircraft was seen to crash into a hill. Lt Carmichael at first thought it was one of the Sea Furies, but a quick 'call of the roll' confirmed it was in fact one of the MiGs. Although Carmichael was credited with the victory, his flight members (Lt Davis and Sub Lts Haines and Ellis) considered they deserved a quarter of a MiG each!

Variants and Exports

Several other Sea Fury variants were produced, among them 61 T Mk.20 two seat conversion and transition trainers for the Royal Navy, created by installing a rear cockpit for the instructor with full dual controls. The front cockpit remained standard and a perspex connecting 'tunnel' joined the two otherwise separate canopies.

Other modifications required included the deletion of one pair of cannon (the space being used to house equipment relocated from the rear fuselage) and of the tailhook. The tailwheel was non retractable and a periscope was fitted above the rear cockpit gunsight so as to allow the instructor to sight over his pupil's head. Underwing hard points were retained.

The first Sea Fury T.20 (VX818) flew in January 1948 and deliveries began in early 1950, the aircraft subsequently being used for a variety of roles apart from conversion training, including communications and instrument flying training.

While Hawker was developing the T.20, it was working to fill a similar requirement from Iraq, which in December 1946 had ordered 30 Sea Fury FB.11s and four trainers. Most of these were delivered in 1948-50 (as Sea Fury F.60s and T.61s, respectively), and were followed by a further 25 single seaters and three trainers from May 1952 after delays caused by the Royal Navy's need for the aircraft.

Two of the original four Iraqi T.61s were not delivered at all, one being diverted to Pakistan and the other becoming the prototype T.20.

These 'Baghdad Furies' as they were known, were denavalised, lacking catapult launch hooks, tailhooks and other naval equipment. The T.61s lacked the cockpit connecting tunnel of their British counterparts and instead featured completely separate canopies. The recovery of 26 Iraqi Furies in the late 1970s has formed the basis of the type's current popularity in the 'Warbird' movement around the world.

The Netherlands, Pakistan and Egypt also ordered new production Sea Furies from Hawker. The Dutch received 22 Sea Fury FB.50s in 1948 and a further 25 examples were built under licence by Fokker for delivery to the Royal Netherlands Navy in 1950-51. Dutch Sea Furies flew from the carrier *Karel Doorman*.

Pakistan took delivery of 98 Sea Furies between 1950 and 1954 including 86 new build FB.60s, the second Fury prototype (NX802) which was brought up to FB.60 standards, five refurbished FB.11s, five new T.61s and a T.61 originally intended for Iraq. Pakistan's Furies lacked naval equipment and operated only from dry land.

Other export customers (apart from Australia and Canada) received refurbished ex Royal Navy Sea Fury FB.11s at various times during the 1950s: Egypt (12 plus the original Fury prototype, NX798); Burma (18 FB.11s and three T.20s); Cuba (15 FB.11s and two T.20s) and West Germany, which had 16 T.20s and an ex Dutch FB.50 operated on its behalf by the company *Deutsche Luftfahrt Beratungsdienst* which was contracted to provide target towing services. These aircraft were flown in civil markings and operated through the 1960s and into the 1970s.

The Cuban aircraft had an interesting history. Ordered by the Batista government in 1957, they were delivered in 1958, the year before the communists under Fidel Castro had seized power. There are no reports of the aircraft being used in the revolt, although they did end up in the communists' service, albeit hampered by a lack of maintenance expertise due to the forced removal of many qualified engineers when the new regime was established.

It is known that five of the aircraft were flown in combat during the US supported Bay of Pigs landings in 1961, attacking ships and aircraft involved in the battle.

FAIREY FIREFLY

Fairey Firefly AS.6 WD884 of the Royal Australian Navy Fleet Air Arm. Australia operated 108 Fireflies.

FAIREY FIREFLY

Originally designed as a two seat naval fighter, the Fairey Firefly evolved into an extremely useful multirole aircraft, setting a standard of versatility which had not previously been approached.

Although the Firefly's wartime career was relatively short (it entered service in late 1943 and didn't see its first action until July 1944), it nevertheless made an impact, its best known actions including photographic reconnaissance of the German capital ship *Tirpitz* in Altenfjord, Norway, immediately before its sinking, and rocket attacks on Japanese held oil refineries in Sumatra in January 1945.

In addition, a Royal Navy Firefly was the first British aircraft to fly over Tokyo during World War II.

The majority of the Firefly's service was therefore postwar, the type remaining with the Royal Navy until 1956, in the meantime serving with distinction in Korea and being developed to operate not only as a naval fighter but also as a reconnaissance, night fighter, trainer, anti submarine, target towing, training, pilotless drone and ground attack aircraft.

Its usefulness is indicated by the length of time it remained in production after World War II. The first Firefly was flown in December 1941, and by the time the final example was delivered in May 1955, a total of 1,701 had been manufactured in seven major variants.

Only two nations operated the Firefly apart from Britain – The Netherlands (40 FR Mk.4s) and the Royal Australian Navy, which received 108 aircraft between 1949 and 1953. Alongside the Hawker Sea Fury (with which the Firefly shared the privilege of being the first equipment of the newly established RAN Fleet Air Arm in 1948), the Firefly operated successfully in Korea, flying from the carrier HMAS *Sydney*.

Flycatcher to Firefly

The Fairey Aviation Company of Hayes, Middlesex, had had a lengthy association with the Royal Navy by the time the Firefly came along. Starting with the 'III' series of fighter/reconnaissance aircraft which first appeared at the end of World War I and continued in production until the early 1930s, the company subsequently introduced the Flycatcher fighter and the immortal Swordfish biplane torpedo bomber in the interwar years.

The Swordfish, of course, remained in production until 1944, outlasting its successor, the Albacore, which first flew in 1938 and was removed from production five years later. The monoplane Barracuda torpedo/dive bomber was built between 1940 and 1946.

The company had in the meantime produced the Fulmar, a two seat monoplane carrier fighter/bomber powered by a single Rolls-Royce Merlin engine and in effect a lighter, refined development of the Royal Air Force's unsuccessful Battle light bomber.

First flown in 1937 to meet an RAF requirement, the basic design was revised to meet a naval requirement and Fulmar became the Royal Navy's first eight gun fighter, entering service in 1940. Its performance (top speed 280mph/450km/h) proved to be adequate if not spectacular but other virtues such as its manoeuvrability, endurance and armament of eight 0.303in Browning machine guns (the same as in early model Spitfires and Hurricanes) meant it was able to hold its own against most opposition encountered at sea.

At its peak in mid 1942, Fulmars equipped 14 RN Fleet Air Arm squadrons and the type served in the European, Mediterra-

A flight of Royal Navy Firefly FR.4s poses for the camera.

Probably the best known product of the Fairey Aviation Company, the Swordfish torpedo bomber.

nean, Middle East and Far East theatres of operation, as well as from CAM (catapult armed merchant) ships. Fulmar production reached 600 examples, the last of them delivered in early 1943.

The design of what would become the Firefly evolved ostensibly from a Royal Navy requirement for a Fulmar replacement, although this is perhaps over simplifying the situation. While intended first and foremost as a naval fighter, the Firefly was also thought of as having multirole capability right from the start.

The British Admiralty formulated new requirements for shipboard fighters in 1939, and as a result issued Specifications N.8/39 and N.9/39, the former for a conventional fixed gun fighter and the latter for a turret fighter.

In view of the lack of success of Britain's existing turret fighters – the RAF's Boulton Paul Defiant and the RN's Blackburn Roc, it was probably just as well that Specification N.9/39 disappeared and a new one – N.5/40 – issued in June 1940 covering a fixed gun fighter and based on Fairey's submissions.

The Fairey designs had been prepared under the leadership of H E Chaplin, who, like the powers-that-be, had concluded a more conventional fixed gun two seat fighter was the way to go.

With the issue of the new specification came an initial contract to Fairey for 200 aircraft, just nine months after the company had submitted its first proposals to the Admiralty in September 1939.

Progress was rapid from there. First drawings were issued in November 1940 and the aircraft was officially named 'Firefly' the following month.

The first development aircraft (Z1826) was flown for the first time from the Great West airfield by Fairey's chief test pilot, C S 'Chris' Staniland, on 22 December 1941, just 13 months after the issue of the first drawings. This and the following three Fireflies were built in Fairey's experimental shop

A Royal Navy Firefly I displays its wing folding method aboard a British carrier. (via Neil Mackenzie)

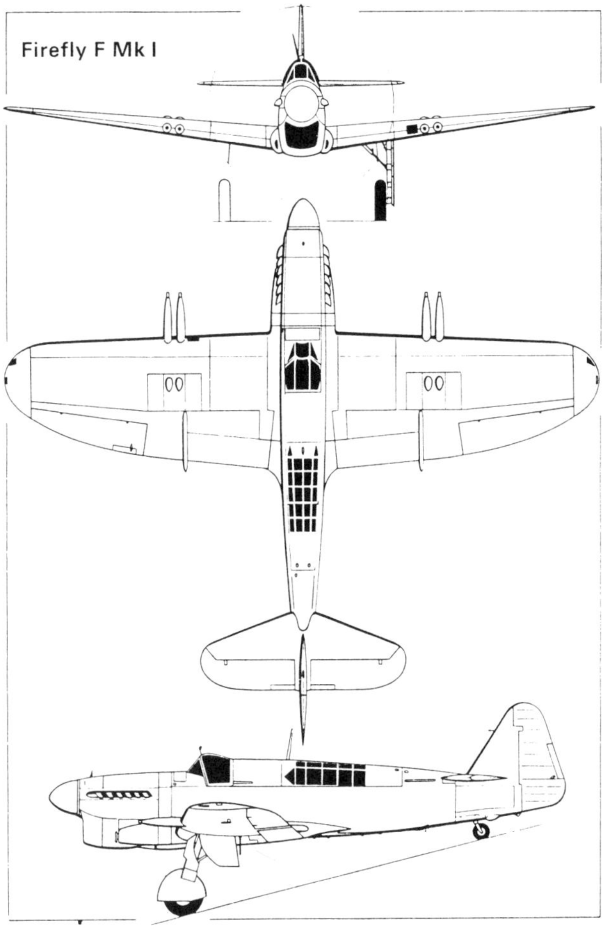

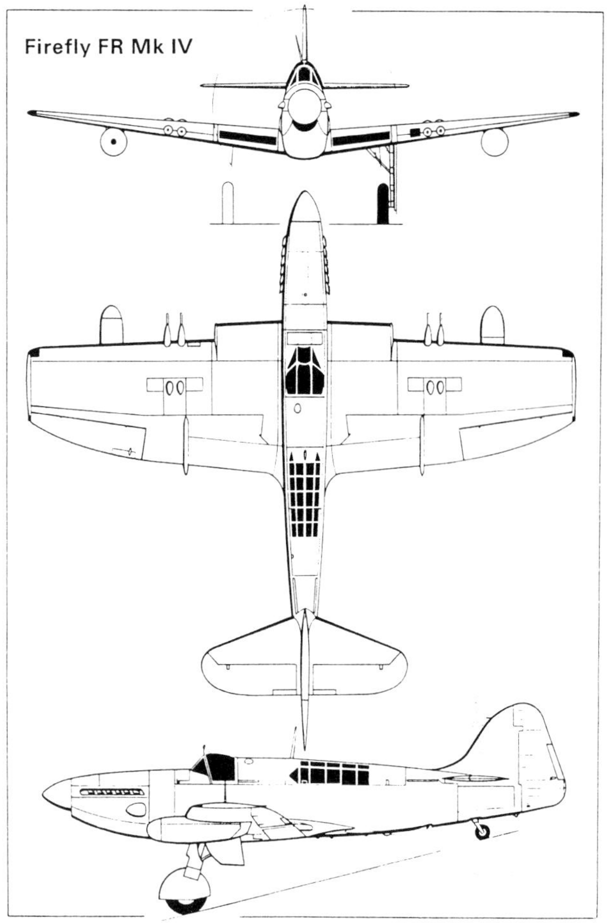

A 1947 photograph of a Firefly I in its natural element. Note the 'beard' type radiator of this early model. (via Neil Mackenzie)

Firefly I Z2118 after conversion to the prototype FR.4 with wing root radiators, clipped wingtips, fin fillet and the outer wing radar and fuel pods.

at Hayes from production components, so there was no prototype in the true sense.

The other three development aircraft (Z1827-29), first flew in March, August and September 1942, respectively. Z1827 was an early casualty, shedding its tail assembly at low altitude just three weeks after first flight following elevator overbalance. The aircraft was destroyed in the crash and Chris Staniland lost his life. The fourth aircraft (Z1829) was originally intended as a structural test specimen but was completed for flight testing as a result of the accident.

Firefly Described

Of all metal construction, the Firefly featured a semi-monocoque fuselage with the tailplane and fin mounted on tubular steel structure at the rear fuselage. All covering was metal with the exception of the fabric skinned rudder.

Power was provided by various marks of the Rolls-Royce Griffon 60 degree vee, 12 cylinder water cooled engine, in the case of early Firefly Is the Griffon IIB rated at 1,720hp for takeoff, 1,735hp at 1,000 feet and 1,490hp at 14,000 feet. The engine was fitted with a two speed/single stage supercharger and drove a Rotol three bladed constant-speed propeller through 0.451 to 1 reduction gearing. The Griffon was mounted as a quickly detachable 'power egg' module and the Firefly was one of the first British aircraft with this feature.

The semi elliptical wing comprised a one piece centre section, 'back to back' spars and stressed metal skin covering. The manually (in early models) folding wings turned upward around rear spar universal hinges and swung backwards to lie along the fuselage sides with trailing edge down. In flight, the wings were locked down hydraulically.

The main fuel tank of 146imp gal (664 litres) capacity was housed immediately aft of the pilot in the centre fuselage and forward of the observer's compartment. A further 46imp gal (209 l) was housed in centre section tanks, while later versions carried fuel in a fixed tank under the port wing.

A notable feature of the wing design was the use of Fairey-Youngmann area increasing flaps, a major contributor to the Firefly's excellent manoeuvrability and low speed handling characteristics. The flaps extended aft on tracks and in addition to providing high lift for takeoff and lift in combination with high drag for landing, had an intermediate setting for use in the cruise and when manoeuvring, the result being a notable reduction in turning radius.

Fixed armament comprised four 20mm Hispano cannon in the wings, each with 160 rounds of ammunition. Attack ordnance was a choice of sixteen 60lb (27kg) rockets or two 1,000lb (454kg) or 500lb (227kg) bombs initially, to which were later added sonobuoys, depth charges, mines and air/sea rescue capsules in the anti submarine variants which followed.

FAIREY FIREFLY FR Mk.4
Powerplant: *One Rolls-Royce Griffon 74 liquid cooled, 12 cylinder vee piston engine with two speed/two stage supercharger rated at 2,020hp for t-o, 2,245hp at 9,250 feet and 1,420hp at 20,500ft. Rotol four bladed constant speed propeller. Fuel capacity 146imp gal (661 l) in fuselage tank and 55imp gal (250 l) in port wing pod; provision for two 90imp gal (409 l) drop tanks.*
Dimensions: *Wing span 41ft 2in (12.54m); width wings folded 16ft 0in (4.88m); length 37ft 11in (11.56m); height 14ft 4in (4.37m); wing area 330sq ft (30.65sq m).*
Weights: *Loaded (clean) 13,500lb (6,123kg); max t-o 15,615lb (7,083kg); empty 9,859lb (4,472kg).*
Armament: *Four 20mm Hispano cannon in wings with 160rpg; provision for two 1,000lb (454kg) bombs or eight 60lb (27kg) rockets or two 500lb (227kg) bombs and eight 25lb (11kg) rockets. Also provision to carry air-sea rescue container, 250lb (1123kg) depth charges or 1,000lb (454kg) mines.*
Performance: *Max speed (clean) 316mph (509km/h) at sea level, 345mph (555km/h) at 12,500ft; Max speed (with drop tanks) 282mph (453km/h) at sea level, 310mph (499km/h) at 12,500ft. Time to 5,000ft 3.6min, to 10,000ft 7.15min, to 20,000ft 15.5min. Service ceiling 29,200ft (8,900m). Range (standard fuel) 582 miles (936km) at 233mph (375km/h); range (two 90imp gal drop tanks) 1,070 miles (1,722km) at 228mph (367km/h).*

RAN Firefly AS.6s in formation. Note the lack of guns on this specialist anti submarine version.

The RAN Historic Flight's Firefly AS.6 WD826 in the air and on the ground, the latter shot giving an unusual angle while the wings fold. (RAN/Bruce Chapman)

Mike Wansey's Firefly WD828 flew for three years after restoration before being severely damaged in a 1987 forced landing. The aircraft was photographed at the 1985 Schofields Air Show. (Bruce Chapman)

Externally, the Firefly I featured an undernose 'beard' style radiator but versions with two speed/two stage Griffons installed (from the Mk.IV onwards with the exception of the final production model) had clean noses and radiators mounted in wing centre section extensions.

RAN Firefly AS.5 VX385 (with guns removed) photographed during a visit to NAS Brooklyn in the USA. (via Mike Kerr)

Testing and Into Service

Service testing of the Firefly began in the second half of 1942, by which time preparations for mass production were well underway. The orderbook was also filling rapidly, and by the time deliveries had begun it had increased to 800 aircraft – 500 Mk.Is, 200 Mk.IIs and 100 Mk.IIIs.

Carrier trials aboard HMS *Illustrious* started late in 1942 and revealed a deficiency in forward view during the approach and a problem with the pilot's canopy, which during one test detached itself, wrapped itself around the Firefly's port tailplane and severed the mainspar. Captain Eric 'Winkle' Brown completed a dodgy circuit to recover to the carrier at high speed despite the damage which included violent stick flutter from the blanked off port elevator.

The incident indicated a canopy jettisoning problem and this in combination with the poor forward view resulted in the fitting of a taller windscreen with thinner frames plus a taller canopy, a bonus for larger pilots.

Other modifications introduced to the Firefly during its trials period included the replacement of the rudder's horn balance with a mass balance, the replacement of fabric covered ailerons with metal ones, the fitting of fairings to the gun barrels and the replacement of the original two man dinghy in the rear fuselage with individual K-type dinghies.

The first production Firefly F.I (Z1830) was delivered to the Fleet Air Arm in March 1943 and the first operational unit to receive the type was No 1770 Squadron at Yeovilton, in October 1943. Embarked aboard HMS *Indefatigable*, this squadron also took the Firefly into battle for the first time in July 1944 with attacks on the *Tirpitz* in Norwegian waters.

No 1771 Squadron was the next to equip with the new fighter and operating from HMS *Implacable*, began operations in and around Norway in October 1944.

Both squadrons moved to the Far East at the end of 1944 and began operations against the Japanese oil refineries on Sumatra early in the new year. It was during this period that 1770 Squadron's Lt D Levitt shot down a Japanese Nakajima Ki-43 'Oscar' fighter and recorded the Firefly's first 'kill'. The Royal Navy operated four Firefly squadrons as part of its Pacific Fleet in the last year of the war, joining US forces on strikes against Okinawa, the Carolinas and other islands as the Allies moved ever closer to Japan.

Production and Models

Production of the Firefly F.1 amounted to 427 aircraft, of which 132 were completed under sub contract by General Aircraft Ltd (GAL). These were followed by the NF.II night fighter with AI Mk.X radar housed in two radomes on the wing leading edges. Centre of gravity problems associated with the weight of the radar operator's equipment in the rear cockpit necessitated a 15 inch (38cm) 'stretch' of the fuselage forward of the front cockpit.

This had undesirable effects on the aircraft's performance and since the modified fuselage removed the basic appeal of the aircraft – a simple conversion of Mk.I aircraft on the production line – production was limited to 37 aircraft.

Instead, the fitting of more compact radar resulted in the Firefly FR.1 fighter/reconnaissance and NF.I night fighter versions, equipped with ASH (ship/submarine detection) and AI (airborne interception) radar, respectively, the latter housed in a pod below the engine.

Firefly FR.1 production reached 236 and NF.I 140 examples, with deliveries completed by November 1946. The NF.IIs were converted back to Mk.I standards, while the majority of FR and NF.Is were fitted with Griffon XII engines rated at 1,765hp.

Other Firefly Mk.I variants were the F.IA (F.Is converted to FR.Is) and two postwar conversions, the T.I dual control trainer and TT.I target tower. The designation Firefly T.2 was given to another postwar trainer development which was similar to the T.I.

The advent of a new series of Griffon engines with two speed/two stage superchargers promised greater power, and to exploit this, the Firefly III was developed. An early production Firefly F.1 was fitted with a 2,035hp Griffon 61 and flown in that form in 1944. Fitted with the original 'beard' radiator, this Firefly III (as it was designated) exhibited aerodynamic characteristics which were considered undesirable and development was abandoned in favour of the Mk.IV with engine radiators moved to

Front view of the Firefly's wing folding mechanism.

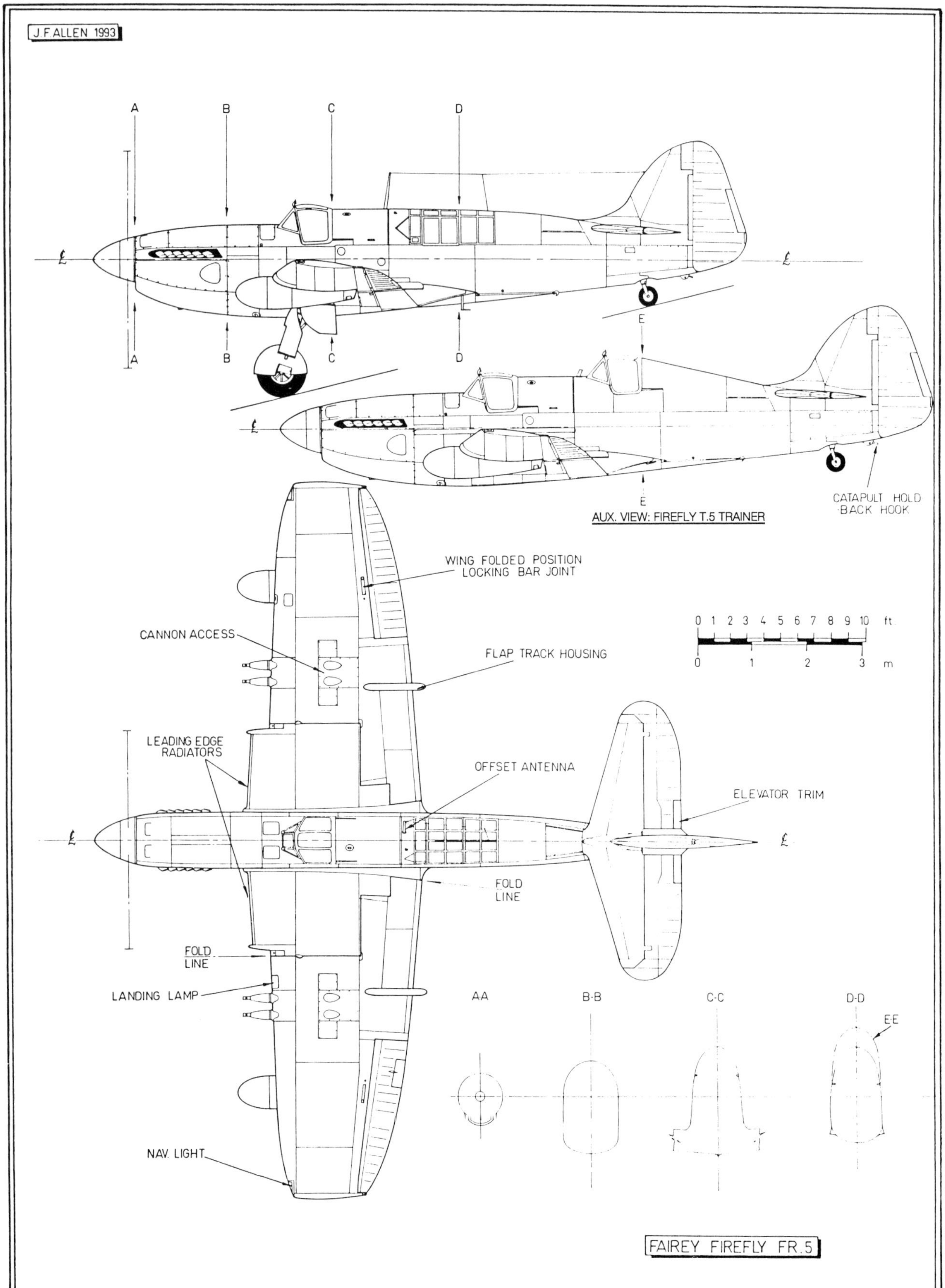
J.F.ALLEN 1993
A
B
C
D
E
AUX. VIEW: FIREFLY T.5 TRAINER
CATAPULT HOLD BACK HOOK
WING FOLDED POSITION LOCKING BAR JOINT
CANNON ACCESS
FLAP TRACK HOUSING
0 1 2 3 4 5 6 7 8 9 10 ft
0 1 2 3 m
LEADING EDGE RADIATORS
OFFSET ANTENNA
ELEVATOR TRIM
FOLD LINE
FOLD LINE
LANDING LAMP
AA
B-B
C-C
D-D
EE
NAV. LIGHT
FAIREY FIREFLY FR.5

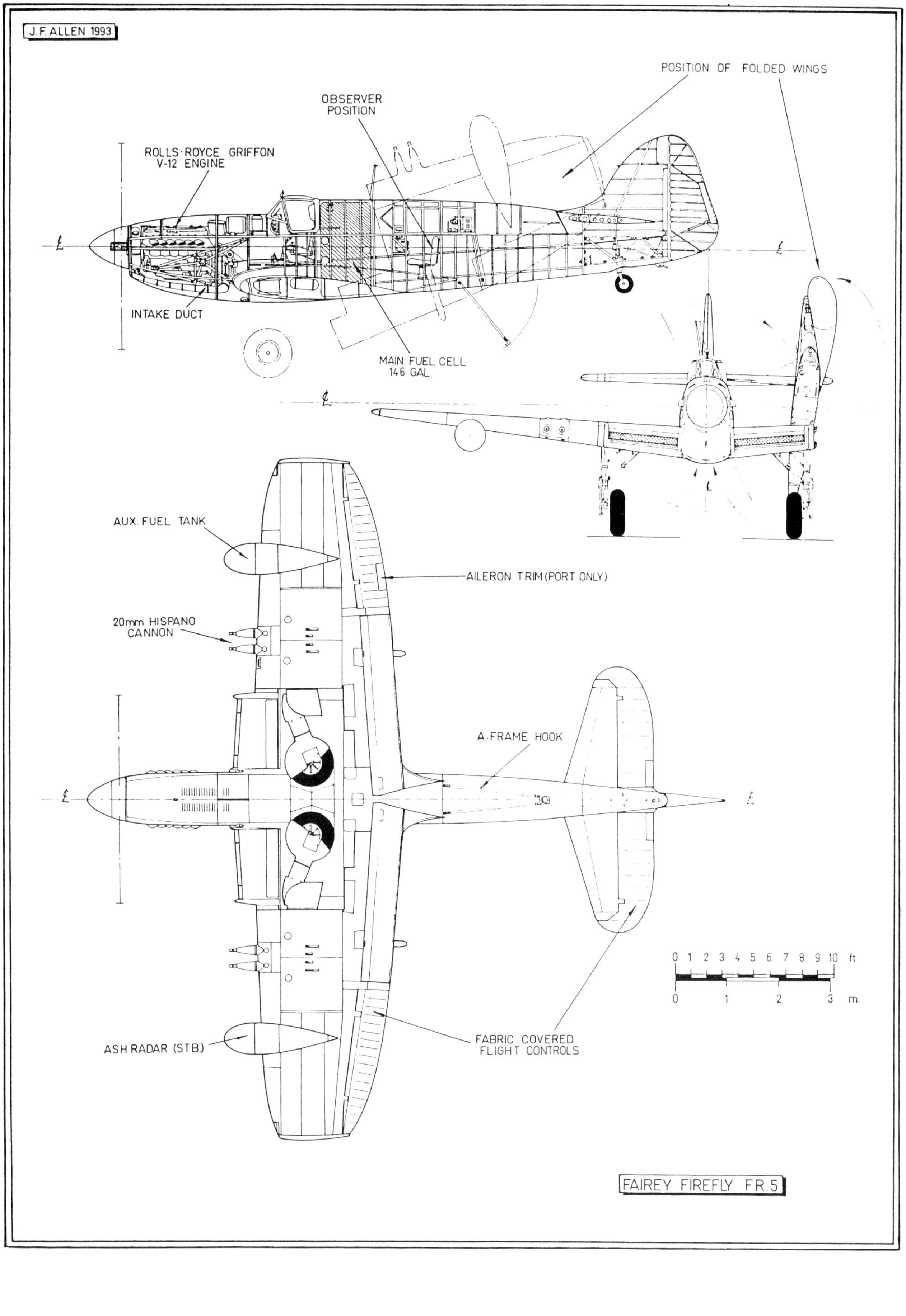

J.F. ALLEN 1993
POSITION OF FOLDED WINGS
OBSERVER POSITION
ROLLS·ROYCE GRIFFON V-12 ENGINE
INTAKE DUCT
MAIN FUEL CELL 146 GAL.
AUX. FUEL TANK
AILERON TRIM (PORT ONLY)
20mm HISPANO CANNON
A-FRAME HOOK
0 1 2 3 4 5 6 7 8 9 10 ft
0 1 2 3 m.
ASH RADAR (STB)
FABRIC COVERED FLIGHT CONTROLS
FAIREY FIREFLY FR.5

(left) The final new production Firefly variant was the much redesigned Firefly AS.7 three seater with full span wings, bulged rear canopy covering the two radar operators, large tail surfaces and a 'beard' radiator. It was developed because of delays to the Fairey Gannet but was basically unsuccessful.

the leading edges of the wing roots, leaving the nose of the aircraft 'clean' and changing its appearance considerably.

Four Firefly Is were converted as prototypes for the Mk.IV from 1944 with the first production example flying in May 1945. Deliveries did not begin until 1947 and production aircraft were powered by 2,245hp Griffon 74 engines driving four bladed Rotol propellers.

The Firefly Mk.4 (British numeric designations changed from Roman to Arabic postwar) also featured clipped wings and an extension of the fin leading edge. Fixed pods under the outer wings housed the radar (starboard) and 55imp gal (250 l) of fuel in the port pod.

Firefly Mk.4 variants were the FR.4 with ASH radar and NF.4 with AI radar plus TT.4 target towing conversions. Production reached 160 aircraft, of which 40 were delivered to the Royal Netherlands Naval Air Service (RNNAS).

The versatility displayed by the Firefly 4 led to further development and resulted in the FR, NF and AS (anti submarine) Mk.5, 352 of which were delivered between January 1948 and May 1950, the prototype having flown in December 1947. Similar to the Mk.4 in its major respects, the Mk.5 introduced one significant improvement over its predecessor with the fitting of a powered wing folding mechanism midway through the production run. Some earlier Mk.5s were retrospectively modified.

A postwar conversion of the Firefly I was the T.1 dual control trainer with a second cockpit for the instructor. These examples are in Dutch service.

Four local Australian trainer conversions of the Firefly 5 were performed by Fairey's Australian subsidiary. The conversion was based on kits supplied from the UK and considerable local redesign as the kits were originally developed for earlier model aircraft. VX373 was the first conversion, performed in 1954. (via Mike Kerr)

The Firefly 4 and 5 were the marks involved in the Korean War, operating from four Royal Navy and one Royal Australian Navy carrier (HMAS *Sydney*) in the close support, minelaying and shipping strike roles, often in conjunction with Hawker Sea Furies. Fireflies flew an intensive sortie rate in Korea – up to 120 per day from a single carrier – and often in very hostile weather conditions.

The specialist anti submarine Firefly AS.6 first flew in March 1949 and entered service the following May, remaining the Royal Navy's main ASW (anti submarine warfare) type until the Fairey Gannet entered service in 1955.

The AS.6 featured British rather than American sonobuoy and associated equipment and had its guns deleted. The 133rd and last Firefly AS.6 was delivered to the Royal Navy in September 1951 and 69 examples ended up in Australian service, joining the two Mk.4s and 37 Mk.5s previously delivered.

The final new production Firefly was the substantially different AS.7, 151 of which were built, ending the type's production run in May 1955. The prototype flew in May 1951.

Developed because of delays being experienced by the Gannet, the AS.7 in the event served with only one FAA squadron and many were converted to T.7 trainers. A three seater, this Firefly looked considerably different from its predecessor thanks to its reversion to a chin radiator, its full span wings, its very much larger fin and rudder and its large bulged rear canopy under which two radar operators sat.

The Firefly AS.7 was powered by a Griffon 59 of 1,990hp with one stage/two speed supercharger and like the AS.6, lacked guns.

The final Firefly variants were conversions: the U.8 was a Mk.7 converted to for use as pilotless drones for use on guided weapons research, while the U.9 was Mks.4 and 5 converted for the same role. The number of conversions totalled 34 and 40, respectively.

Ground running shot of the RAN Historic Flight's Firefly AS.6, WD826. (RAN)

AUSTRALIA'S SEA FURIES AND FIREFLIES

The Australian government's 1947 decision to establish a Fleet Air Arm paved the way – at last – for the purchase of aircraft carriers and the aircraft to operate from them.

As described in the opening chapters, HMAS *Sydney* (commissioned 16 December 1948), HMAS *Vengeance* (on loan from the Royal Navy between November 1952 and October 1955 while HMAS *Melbourne* was being completed) and HMAS *Melbourne*, commissioned in October 1955 after modification to operate jet aircraft, between them served Australia for 34 years.

The nucleus of the Fleet Air Arm was Australian airmen who had served with the British Pacific Fleet during World War II along with former RAAF pilots, the Australians attached to the Royal Navy remaining with that service until the RAN aircraft carriers were in service.

The 1947 Five Year Defence Programme under which the FAA was established estimated the capital cost of two light fleet carriers at just under £12m ($24m) plus maintenance costs of nearly £11.5m ($23m). In September 1947 the Australian government placed an initial order for 90 aircraft to equip the two aircraft carriers.

Ultimately, the RAN would officially take delivery of 101 Sea Furies and 108 Fireflies. All but one of the Sea Furies would be FB.11s (there was a single F.10 officially listed, more of which below) while the 108 Fireflies were divided into two FR.4s, 37 AS.5s and 69 AS.6s. In both cases, additional aircraft which remained the property of the Royal Navy were flown by the Australian squadrons during the Korean conflict.

Working Up

August 28, 1948 could be regarded as the date of the birth of the RAN Fleet Air Arm, for on that day the 20th Carrier Air Group (CAG) was commissioned at RNAS Eglinton, Northern Ireland. The CAG comprised 805 Squadron with Sea Furies and 816 Squadron with Fireflies, under the command of Lt Cdr P E I Bailey and Lt Cdr C R J Coxon, respectively.

An extensive working up period followed, using aircraft lent to the RAN squadrons by the Royal Navy until the 'official' Australian aircraft became available in early 1949.

In the case of the Sea Furies, 16 F.10s and six FB.11s are known to have been used by 805 Squadron at Eglinton during this period. For the record, they were F.10s: TF903, 910, 911, 913, 920, 923, 925, 926, 927, 940, 943, 945, 950, 951, 952 and 953 plus FB.11s: TF956 (the first production Mk.11), 973, 990, 991, 992 and VR950.

Of those, TF926 and TF953 were written off in accidents during the working up period, while some mystery surrounds TF925,

Dramatic view of Bruce Andrews' ex Iraqi Fury flying in RAN colours. This aircraft now resides in Britain with the Old Flying Machine Company. (RAN)

the only Sea Fury F.10 to appear on the list of aircraft officially handed over to the RAN.

TF925 crashed at Eglinton in early 1949 whilst on loan to the RAN and is recorded as having been scrapped in November of that year. It is also recorded as having been delivered to Australia aboard HMAS *Sydney* on the carrier's first trip home after commissioning, along with the other Sea Furies and Fireflies which were carried.

It is perhaps reasonable to suggest that this Sea Fury was transferred to the RAN after the accident, but why it would be recorded as having been delivered to Australia remains something of a mystery. Regardless, TF925 was definitely struck off the RAN's books in November 1949.

Other Sea Furies and Fireflies were lent to the RAN during the working up period for 808 and 817 Squadrons in 1950, immediately before HMAS *Sydney's* second delivery trip between the United Kingdom and Australia near the end of that year. Once again for the record the Sea Fury FB.11s known to be involved were VW652, 654, 663, 669, 692, 712 and 717, VX608, 613, 619, 633, 657, 659 and 692, a total of 14 aircraft.

Other Sea Furies which were operated by the RAN but did not officially appear on strength were those lent by the Royal Navy during HMAS *Sydney's* period of operation in Korea. These are separate from the attrition replacement aircraft (see below) which were taken on charge by the RAN.

Five aircraft come under this category – WE725, 786, 790 and 805 plus WF592 – all of which served aboard HMAS *Sydney* during its more than three months of active service in Korea during 1951-52.

To further add to the list of 'unofficial' RAN Sea Fury FB.11s, six others were loaned to the RAN in early 1954 and were allocated to 850 Squadron aboard HMAS *Sydney*. They were WG597, WJ226, WJ241, WJ246, WJ286 and WJ296 and all were provided from RN stocks at Iwakuni, Japan while *Sydney* was undertaking her second voyage in Korean waters, this time in a peacekeeping role.

Firefly AS.6s overfly HMAS Sydney *with WD837 in the foreground. (via Mike Kerr)*

Deliveries

After working up in the United Kingdom, the squadrons embarked on the newly commissioned HMAS *Sydney* and sailed for Australia on 12 April 1949, arriving at Jervis Bay on 25 May. On board *Sydney* for her maiden voyage were 56 aircraft, including 27 Sea Furies and 26 Fireflies, of which two were FR.4s and the remainder AS.5s.

HMAS *Sydney* returned to the United Kingdom in November 1950 to collect a second batch of aircraft for the squadrons (808 with Sea Furies and 817 with Fireflies) which would make up the 21st Carrier Air Group. This time the voyage to Australia would see the carrier with 32 Sea Fury FB.11s and a similar number of Fireflies on board, two of them AS.5s and the remainder the specialist anti submarine AS.6. *Sydney* arrived in Australia in early December 1950 with her new aircraft.

In the meantime, two Furies and two Fireflies had arrived in Australia by other means. One of each was delivered by merchant vessel aboard the SS *Sussex* in March 1950, while another pair arrived two months later aboard the SS *Stentor*.

HMAS *Sydney's* involvement in the Korean war between October 1951 and January 1952 brought with it the need to replace aircraft lost in action, these coming from Royal Navy stocks and provided via the Royal Navy carrier HMS *Unicorn*, which acted as a resupply, repair and occasional troop ship during the conflict, operating out of Hong Kong. During the Sydney Air Group's period of active service in Korea, three replacement Firefly AS.5s and eight Sea Furies were transferred from *Unicorn*.

Fireflies of the first delivery batch aboard HMAS Sydney *whilst still in the United Kingdom. (RAN)*

One of the first of the RAN's Sea Furies, VW647, during unloading upon arrival in Australia. (RAN)

HMAS Vengeance photographed in Fremantle in 1953 after returning from the UK with a load of Fireflies and Sea Furies on board. (RAN)

Fireflies and crews from HMAS Sydney *at the Naval Air Station, Brooklyn USA on the way back from the ship's Coronation voyage of 1953. Pictured at Floyd Bennett Field are (left to right): Lt Cdr D S Harvey, Lt H E Bailey, Sub Lt D M Kimber, Comm Engr G A McGregor, Sub Lt A B Haywood, Lt Cdr J E Gillow and Lt D J Robertson. (via Mike Kerr)*

Korean activities resulted in an almost unnoticed but significant change of equipment for 817 Squadron. Originally formed with Firefly AS.6s, the decision to send the squadron to Korea meant this Firefly variant had to be 'swapped' for 816 Squadron's AS.5s because the later model did not have any internal gun armament. The AS.5's four 20mm cannon were considered (and proven to be) necessary for the types of operations which would be undertaken in Korea and a partial exchange of equipment was affected.

Only six Firefly 5s went to Korea with 817 Squadron and many extra aircraft were provided by the Royal Navy. These were never officially taken on RAN charge and were returned to the British. Royal Navy Sea Furies also served with the Australian squadrons during HMAS *Sydney's* time in Korean waters.

It is known that there were experiments conducted with a view to fitting guns into the wings of a Firefly AS.6, but these were apparently abandoned.

Deliveries of Sea Furies and Fireflies continued after the end of the Korean conflict, *Sydney* bringing batches of aircraft to Australia in March 1952 (10 Sea Furies and four Fireflies from NAS Sembawang, Singapore), August 1953 (18 Firefly AS.6s from the UK) and January 1954 (12 Sea Furies transferred from HMS *Perseus*).

HMAS *Vengeance* was also used to transport aircraft to Australia during her brief career with the RAN. Following her commissioning into RAN service in November 1952 and the completion of a refit the following January, *Vengeance* sailed for Australia with 10 Sea Furies and 20 Firefly AS.6s (plus three Bristol Sycamore helicopters) on board. The Sea Furies and Fireflies were to equip newly recommissioned squadrons, 850 and 851, respectively, the former serving aboard *Vengeance* with 805 and 816 Squadrons in preparation for a stint of Korean war service which never materialised. 851 Squadron served as a training unit for pilots and observers and was based at HMAS *Albatross* at Nowra.

Sea Fury Squadrons

Three 'front line' RAN squadrons operated the Sea Fury. Prior to reforming as an Australian unit, 805 Squadron was a Royal Navy unit which had seen service in the Western Desert between 1940 and 1942 flying Fairey Fulmars, Brewster Buffalos and Grumman Martletts. It also operated in Crete (with some RAF Hurricanes) and it was there the squadron had the misfortune of losing all its aircraft in two days.

Disbanded at Nairobi in January 1943, 805 Squadron was reformed in July 1945 with Seafire IIIs with the intention of operating in the Pacific, but the sudden end of the conflict put paid to those plans although it remained active until 1948.

The squadron was recommissioned as a Royal Australian Navy unit in August 1948, equipping with Sea Furies and becoming part of the 20th Carrier Air Group aboard HMAS *Sydney*. 805 Squadron flew in the Korean War (refer the following chapter) and was the last RAN Fleet Air Arm front line squadron to operate the Sea Fury, swapping them for Sea Venoms in 1958.

808 Squadron was another former Royal Navy unit, spending the last three years of World War II operating from escort carriers. It was disbanded in December 1945 but reformed in the United Kingdom in April 1950 as the second RAN Sea Fury squadron, forming part of the 21st Carrier Air Group. 808 Squadron also fought in the Korean War and continued operating from HMAS *Sydney* until September 1953 when it was transferred to HMAS *Vengeance*. The squadron was the first to relinquish its Sea Furies when it disbanded in October 1954. It reformed the following year, re-equipped with Sea Venoms.

The Sea Fury was officially retired from front line service on 18 March 1958 when 805 Squadron made a farewell flight over Sydney Harbour.

850 Squadron was also a former Royal Navy unit before becoming part of the RAN, but had only a brief period of Australian service. Reformed at Nowra in January 1953, the squadron's Sea Furies served aboard both HMAS *Sydney* and *Vengeance* before being disbanded in August 1954. In the meantime, the squadron and its aircraft did see some Korean service as part of HMAS *Sydney's* peacekeeping patrols in the area between November 1953 and June 1954.

Firefly AS.5 VX385 at NAS Brooklyn, USA, in 1953. (via Mike Kerr)

(left) With Fireflies and Sea Furies ranged aft, HMAS Sydney *'dresses up' for an arrival into port.*

Firefly Squadrons

816 Squadron enjoyed an active service life with the Royal Navy during World War II before being commissioned as an RAN unit at Eglinton, Northern Island in August 1948 as part of the 20th Carrier Air Group and equipped with Firefly AS.5s.

The squadron first operated Fairey Swordfish torpedo aircraft from HMS *Furious* on the Atlantic convoys in 1939 before serving in the North Sea off Norway on reconnaissance and anti submarine patrols and participating in torpedo attacks on shipping at Trondjhem.

816 Squadron was aboard HMS *Ark Royal* when she was torpedoed and sunk in November 1941. The squadron reformed in Jamaica in February 1942 and again equipped with Swordfish (as well as Wildcats and Seafires later on) served aboard several carriers. In 1944, whilst escorting Russian convoys, 816 was credited with the destruction of four U-boats.

The years 1945-48 saw the squadron equipped with Fairey Barracudas and Firefly Is and operations were conducted in the Mediterranean before disbandment and then reformation as an RAN unit. 816 operated from HMAS *Sydney* until the ship was sent to Korea in late 1951 and for four months from April 1953 joined 805 and 850 (Sea Fury) Squadrons aboard HMAS *Vengeance* for working up trials in preparation for a tour of duty as part of the peace keeping efforts in Korea following the uneasy truce there.

(below) A spot of bother for a Sea Fury pilot who has caught one of Sydney's *wires successfully but then veered to port. The photograph is dated April 1956. (RAN)*

Vengeance was not sent to Korea in that capacity, HMAS *Sydney* going instead for six months from November 1953 with 816 Squadron's Fireflies aboard. The squadron was disbanded at Nowra in April 1955 and reformed four months later in the United Kingdom, re-equipped with Fairey Gannets.

817 Squadron was another unit with a fine war record whilst part of the Royal Navy, winning battle honours in Norway in 1941 plus North Africa and Biscay after that. The squadron disbanded at the end of World War II to be recommissioned in April 1950 as an RAN squadron equipped with Fireflies and initially operating as part of the 21st Carrier Air Group in conjunction with 808 Squadron's Sea Furies.

As discussed in more detail in the following chapter, 817's Fireflies served with 805 and 808 Squadrons' Sea Furies aboard HMAS *Sydney* during the Korean War. The squadron was embarked on HMAS *Vengeance* between September 1953 and April 1955 when it was disbanded in preparation for reforming with new equipment, the Fairey Gannet.

Firefly AS.5 VX388 sporting the 'Q' fin letter which identifies it as operating from HMAS Vengeance *at the time. (via Mike Kerr)*

817 Squadron's Fireflies took part in the April 1953 Coronation Review Flypast at Spithead. The RAN had been invited to participate and HMAS *Sydney* was sent for the occasion, carrying the 'Coronation Contingent', a group of representatives from all three Australian services. A total of 17 Firefly squadrons participated in the flypast, including eight aircraft from 817 Squadron.

A third RAN squadron operated a majority of Fireflies, although in this case its role was training and it was shore based at Nowra. 851 Squadron had also originally been a Royal Navy unit having disbanded in 1945. It was reformed at Nowra in August 1954 equipped with 16 Fireflies and two Dakotas and provided training for pilots and observers. It was disbanded in January 1958.

Mixed Squadrons

Several Fleet Air Arm squadrons were commissioned in the 1950s as training and miscellaneous units, usually operating mixed and ever changing fleets of aircraft. Like the operational squadrons, these too had their origins in Royal Navy units.

723 Squadron was reformed at Nowra in April 1952 equipped with two Sea Furies, a Dakota, a Wirraway and a Sea Otter, its roles including fleet support, communications, search and rescue, refresher flying and instrument training.

Before temporarily disbanding in October 1956, Firefly Trainers (see below), Sycamore helicopters and Vampire Trainers were

(left) Sea Fury VW643 of 805 Squadron in the act of coming to grief (top) and after doing so aboard HMAS Sydney *in February 1950. Although of poor quality, the photographs are of interest. (Les Bushell)*

Another one does a 'barrier', this time Sea Fury VW624 aboard HMAS Sydney.

Aircraft carrier operations has its problems, as this Firefly crew discovers as the aircraft slides into the barrier minus its starboard undercarriage leg. (RAN)

added, some of the squadron's civilian rescue and flood relief activities in the meantime attracting public attention.

723 Squadron was reformed in February 1957 with Sycamores and Firefly target tugs for Fleet Requirements Unit (FRU) operations. By 1963 the squadron was an exclusively rotary wing unit.

724 Squadron's history is also described in the Sea Venom section of this book, but in summary it was commissioned at Nowra in June 1955 as an operational flying school and to convert Sea Fury and Firefly pilots to the Sea Venom and Gannet.

724's Sea Furies and Fireflies were transferred to 805 and 851 Squadrons, respectively, in October 1956, these two types returning nearly five years later with the disbandment of 725 Squadron (see below). The squadron's role was by now that of training and providing fixed wing fleet requirements, the elderly piston engined aircraft being used for the latter role, the Fireflies having been converted to target tugs.

By then, the Sea Furies and Fireflies were well into the twilights of their careers, and it was with 724 Squadron that they ended their flying days with the Royal Australian Navy. The Sea Fury was used until October 1962 while the last RAN Firefly flight was recorded on 1 March 1966.

The third of the 'miscellaneous' FAA squadrons was 725, commissioned at Nowra in January 1958 as a Fleet Requirements Unit (FRU). Its initial mixed fleet comprised one Dakota, two Austers, three Gannets, one Sea Fury and two Fireflies, one of which was a target tug.

In May 1959, 725 Squadron's role was changed to that of anti submarine warfare training but the fleet remained mixed and now included two Fireflies and three Sea Furies. The squadron was disbanded May 1962 and its aircraft absorbed by other units, the Sea Furies and Fireflies going to 724 Squadron.

Trainers and Tugs

The need for a Firefly conversion trainer arose in the early 1950s with the result that the decision was made to investigate the feasibility of converting several Firefly Mk.5s to full dual control aircraft, with a raised second cockpit for an instructor where the observer's station was normally located.

The British had previously performed trainer conversions on early model Fireflies and conversion kits were available, but the conversion of the later Firefly 5 necessitated considerable modification.

The work was performed by the Fairey Aviation Company of Australasia, based at Bankstown NSW, using kits supplied by its parent company in the UK but incorporating a substantial amount of local redesign.

Four conversions of RAN Fireflies were carried out, the first of them (VX373) being completed in 1954. The other aircraft involved in the programme were VT440, VT502 and VX385, with the final conversion being performed entirely from a locally designed and manufactured kit.

The recovery of Firefly AS.6 WB521 after a landing accident aboard HMAS Sydney. *The aircraft was converted to components in 1954. (Keith Western)*

Mike Wansey's 'Mickey's Mouse' rotates from NAS Nowra to entertain yet another large air show audience. Unfortunately the aircraft was severely damaged in late 1987 and is unlikely to fly again. (bottom) One of Australia's mini air force of surviving warbirds is Bruce Andrew's immaculate Fury, seen here over the Nowra countryside in 1986. (Eric Allen & Navy PR)

Another Fury flying in the South Pacific region is New Zealand's 'WJ232' (top), another former Iraqi aircraft but fitted with folding wings from a Sea Fury. It is photographed at Wanaka in 1992. The RAN Historic Flight's Firefly is illustrated below. (Bruce Chapman)

Aspects on Fury 'Magnificent Obsession', another former Australian warbird which now flies overseas. In the upper shot it shares the flightline with impressive company, including two other Furies. (Bruce Chapman)

The RAN Historic Flight's WD826 was the only airworthy Firefly in Australia in 1993. (Bruce Chapman/David Wilson)

The lowdown on Guido Zuccoli's Fury and the RAN Historic Flight's Firefly. (Keith Myers/Bruce Chapman)

After conversion, the aircraft were redesignated Firefly T.5s.

Other RAN Fireflies were converted to target tugs for fleet support duties. The conversion involved a kit supplied by Fairey which had been developed for Royal Navy Firefly target tugs. The Type G Mk.3 cable winch and its associated equipment was carried in a nacelle underneath the aircraft, driven by a feathering windmill propeller.

Two Firefly 5s (VX388 and WB271) plus six Firefly 6s (WB518 plus WD826, 828, 840 and 901 along with WJ109) were converted to target tugs for the RAN. After conversion, the aircraft were redesignated Firefly TT.5 and TT.6.

The Fleet Air Arm, 1955

1955 could be regarded as the year the RAN Fleet Air Arm was at some kind of peak with two aircraft carriers in commission, two air stations – HMAS *Albatross* at Nowra NSW and HMAS *Nirimba* at Schofields west of Sydney – plus the satellite airfield at Jervis Bay near Nowra. The FAA had about 80 aircraft on strength at the time.

It was also the period immediately before the Sea Fury and Firefly began to be replaced by the Sea Venom and Gannet in front line service.

The carriers in service in 1955 were of course HMAShips *Sydney* and *Vengeance*, the former being used in an operational role and the latter for training. Squadron strength comprised five front line (with Sea Furies and Venoms) and two second line units.

The philosophy governing the use of the squadrons revolved around them being moved not in groups but as separate entities in order to give greater flexibility. The Flag Officer-In-Charge, East Australian area, administered the second line squadrons and the disembarked front line squadrons in conjunction with the Flag Officer Commanding, HM Australian Fleet, who was the administrative authority for embarked front line squadrons.

All second line squadrons and disembarked front line squadrons were based at Nowra, at that time commanded by Capt P E Fanshawe OBE DSC RN. HMAS *Sydney* was at then under the command of Capt G C Oldham DSC RAN.

Front line operational aircraft were the Sea Fury for interceptor fighter and close support duties and the Firefly AS.6 in the anti submarine role.

The active Fleet Air Arm squadrons in mid 1955 were as follows:

723 Squadron (Lt Cdr J G B Campbell DFC, commanding) – communications, fleet support, search and rescue, training and miscellaneous duties equipped with Sea Furies, Fireflies (including trainers), Vampire Trainers, Sycamore helicopters.

724 Squadron (Lt Cdr L A Robinson) – newly commissioned to carry out operational flying training and to convert Sea Fury and Firefly pilots to the forthcoming Sea Venom and Gannet.

805 Squadron (Lt Cdr F T Sherborne) – fighter squadron equipped with Sea Furies.

Sea Fury VX753 with HMAS Vengeance's *'Q' code on the tail. As became more common as the 1950s wore on, this aircraft has its pilot's name below the canopy. (via Mike Kerr)*

Firefly 5 VX385 of 817 Squadron over Kingston, Jamaica in 1953 during one of HMAS Sydney's *cruises. This aircraft was converted to a dual control trainer shortly afterwards. (L Kenderdine)*

A perfectly judged approach and flare results in a good landing by the pilot of this Firefly 6 aboard HMAS Sydney.

(left) Sea Fury WH589 in the non standard deep blue painted on a couple of aircraft for flying display purposes.

808 Squadron (Lt Cdr G M Jude) – fighter squadron equipped with Sea Furies.

816 Squadron (Lt Cdr D J Robertson) – anti submarine squadron equipped with Fireflies but in mid 1955 preparing to change over to Gannets.

817 Squadron (Lt Cdr J D Goble) – anti submarine squadron equipped with Fireflies but like 816, preparing to change to Gannets.

851 Squadron (Lt Cdr D C Johns) – pilot and observer training squadron equipped with Dakotas and Fireflies.

The Annual Fleet Programme was mapped out well in advance and for the Sea Fury and Firefly pilots this meant a working up period and then a typically four month cruise. The working up period involved weapons training and the all important deck landing practice and qualification.

Aerodrome Dummy Deck Landings (ADDLS) were carried out at Nowra's satellite airfield at Jervis Bay and not at Nowra itself because of the infamous downdraught created by the local terrain on Runway 26. Deck landing practice (DLP) aboard the carrier involved the need to perform six successful full stop landings, the completion of these allowing the pilot to embark.

A stringent 10 day currency was enforced on deck landings, meaning that requalification was often necessary doing ADDLS. Pilots could expect to fly about 20 hours per month when embarked and around 25 per month when ashore.

The all important batsmen, on whom the pilots relied for accuracy of approach during a deck landing, were themselves pilots who undertook a two year tour of duty performing that role.

Sea Furies on parade. A neat line up of aircraft with drop tanks attached. (via Graeme Sivyer)

Even though the Firefly AS.6 was the standard shipborne anti submarine aircraft in the RAN (and Royal Navy) in the early 1950s, it was really obsolescent even as it entered service with its single operator/observer (plus pilot) and limited airframe capability.

Despite this, it was adequate for Australia's needs until the Gannet came along, although operations could be described as being somewhat 'rough and ready'. Even if the Firefly itself was adequate, its detection equipment was quite often not, the non directional sonobuoys it carried (dropped in a cloverleaf pattern) making the business of detection somewhat chancey. Directional sonobuoys made things more effective.

(left) A crowded flight deck with Fireflies preparing to depart.

Sea Fury versus Auster!

The morning of 30 August 1955 witnessed an extraordinary incident which resulted in probably the only recorded example of an aircraft shot down by the RAN Fleet Air Arm. The incident also provided the navy with something of a public relations coup over the RAAF.

It all started when the pilot of Auster J/4 Archer VH-AET was starting his light aircraft at Sydney's Bankstown Airport by swinging the propeller by hand. As occasionally happens under those circumstances, the Auster's engine came to life and the aircraft began to roll.

But this time it was under full power and promptly took off *sans* anyone at the controls, gradually moving in a north-easterly direction over Sydney's suburbs and climbing all the while.

After 30 minutes had elapsed the rampant Auster had still not received any attention from the RAAF but was being shadowed by another of its ilk, this time an RAN J/5G Autocar. Eventually, a Wirraway was 'scrambled' from RAAF Richmond west of Sydney to shoot down the Auster (the civilian one, that is!) but by then the aircraft had climbed to 10,000 feet and the Wirraway's guns had frozen and jammed.

The single act farce continued with the arrival of an RAAF Meteor F.8 jet fighter from No 75 Squadron at Williamtown, flown by Sqdn Ldr M Holdsworth. Holdsworth moved his Meteor into a firing position but found his guns had also jammed! He had to break off and divert to Sydney Airport due to fuel shortage.

A second Meteor with Sqdn Ldr J H Flemming at the controls then arrived on the scene, ready to have a go at the Auster, but before he could 'join battle', two Sea Furies from 805 Squadron RAN swept in and finished the job.

Flown by Lts J R T Bluett (in VW645) and P F McNay (in WZ650) – both Royal Navy officers on exchange duty – the Auster was summarily despatched to a watery grave off Broken Bay north of Sydney at 11.43am. The two Sea Furies had been engaged in gunnery practice near Nowra when the Auster's antics were reported and in the great tradition of the Senior Service, promptly took charge!

The RAAF later explained that its Williamtown fighter squadrons were on stand-down at the time, so it's probably just as well the Russians hadn't decided to come on that particular morning!

An Auster J/4 Archer of the type shot down by RAN Sea Furies in the famous incident of August 1955. (via Stewart Wilson)

Disposals and Survivors

Disposal of both the Sea Fury and Firefly began in earnest in the late 1950s, the years 1957-59 seeing the removal from charge of many aircraft. Of those disposed of during that time, substantial numbers (63 Fireflies and 26 Sea Furies) were sold to scrap metal dealers (or recyclers as they would call themselves today), companies such as A G Simms, T Carr & Co and Fitzgibbon & Garthon.

The aftermath of a potentially disastrous incident during an air display at Griffith NSW in May 1955. Six RAN Sea Furies were the main attraction of the event, run by the Griffith Aero Club to raise funds. The Sea Furies gave an impressive display culminating in carrier style landings at 30 second intervals. Unfortunately, dust generated by the aircraft landing on the unsealed runway reduced visibility to almost nothing, with the result that the fifth aircraft (WZ642) overran the fourth (WZ652), severing its fuselage midway between the cockpit and tail. Happily, neither pilot was hurt. (Frank Scott)

Rear view of a Firefly 6 photographed at Griffith NSW in May 1955 at the air display at which the two Sea Furies photographed elsewhere came to grief. (Frank Scott)

A Firefly 5 from the first batch delivered prepares for launch aboard HMAS Sydney. *(RAN)*

Wearing an overall natural metal finish and 'kangaroo' roundels, Sea Fury WH588 undergoes and engine run. Note the way the aircraft is tied down at the rear. The aircraft carries 805 Squadron's crest on its nose.

Hooks neatly and uniformly down, Fireflies from 817 Squadron photographed during the Coronation Review Flypast at Spithead in April 1953. (L Kenderdine)

These are noted in the Sea Fury and Firefly tables later in the book, as are those aircraft which ended their days on Nowra's fire dump (seven Fireflies and 12 Furies).

Despite these disposals and another round of them in the 1960s, seven former RAN Sea Furies and 11 Fireflies survived, although three of the latter which were made airworthy have subsequently suffered accidents.

Two of the Sea Furies (VW647 and WG630) had unusual retirements. Both were acquired by the Department of Works Experimental Building Station at North Ryde, NSW in late 1959. They were used in the testing of wall panels, the aircraft rigged up in a structure which enabled them to be run at full power, the propeller wash from the Centaurus engine being used to blast against the panels and test their strength.

VW647 was used initially before being sold to the Camden Museum of Aviation, while WG630 – which had been kept in reserve – took over after that. WG630 was in 1993 in the care of the Naval Aviation Museum at Nowra in company with VW232 (carrying the marks 'VX730') and VW623.

Two other former RAN Sea Furies survive and fly in the USA. WH587 is with Ellsworth Getchell of San Jose, California, while WH588 is with Lloyd A Hamilton of Santa Rosa, California. Hamilton also has major components of WH589 in his Pratt & Whitney R-4360 powered unlimited racer *Furias*.

Some of the 26 former Iraqi Furies purchased by Americans Ed Jurist and David Tallichet in 1979 have had an Australian connection as well. Guido Zuccoli acquired three of these, all of which were airworthy in 1993. Two of these were resold, one to Bruce Andrews, who flew his deep blue VH-HFX/'WH589' in Australia for six years before selling it to Ray and Mark Hanna's Old Flying Machine Company at Duxford, England in 1991.

The second ex Zuccoli Fury (VH-HFA/ '253K') was sold to Ted Allen in 1984 but it found its way to the ownership of the War Eagles Air Museum at Santa Teresa, New Mexico, in 1988.

Guido Zuccoli's own aircraft (VH-HFG/ '308') was the only Sea Fury flying in Australia in 1993, now reconfigured to seat two people under a slightly elongated single canopy.

The fourth 'Baghdad Fury' to find its way to Australia is owned by Western Australia's Rob Poynton, who is restoring the aircraft to fly. When it does, it will carry the civil registration VH-ISS.

Eight former RAN Fireflies survive, two of them airworthy in 1993. The first is WB271 which flies with the Fleet Air Arm Historic

Assorted Fireflies in the mid 1950s (top to bottom): T.5 trainer VX373, Mk.5 VT504, and Mk.6s WD879 and WD887, both with RATOG packs visible above the wing leading edge. All aircraft carry the Nowra identification letters 'NW'. (Richard Hourigan)

(left) The First Sea Lord, Louis Mountbatten, toured Australia in 1956. His Handley Page Hastings transport was escorted at one stage by RAN Sea Furies and this photograph was taken at that time.

'Bombed up' Sea Furies in formation with WH581 in the foreground and WH587 behind. (RAN)

An HMAS Vengeance based Firefly AS.6, WD837, of 817 Squadron in late 1953. Like others of its ilk, this aircraft ended its days on fire fighting practice duties. (L Kenderdine)

Assorted Sea Furies No.1 (top to bottom): WZ645, VW632, VW645 and WZ649. (Richard Hourigan)

Assorted Sea Furies No.2 (top to bottom): WJ279, VW634 under guard after a spot of bother, VX758 and VX730, 'beached' after a forced landing. (Richard Hourigan)

Sea Furies (foreground) and Fireflies at Camden NSW during a display. Sea Fury VW645 at the front of the line was eventually sold for scrap. (Peter Ricketts via Ross Gillett)

Flight at Yeovilton, UK. This Firefly was purchased by tender in 1966 for the princely sum of $400, the money being put up by members of the HMS *Victorious* ship's company and RAN personnel.

Airworthy in Australia is WD826/VH-NVU with the RAN Historic Flight, while the associated Naval Air Museum at Nowra also has WJ109. Other Fireflies are under restoration to airworthiness, WB518 with Classic Aviation at Bankstown NSW and WD833 with Henry Schroeder of Danville, Illinois in the USA.

Australian Fireflies on static display or under restoration for display in 1993 were VX388 at the Camden Museum of Aviation, WD827 at the Moorabbin Air Museum and WD840 at the Lone Star Museum, Galveston, Texas USA.

Three RAN Fireflies which had been restored to airworthiness have unfortunately come to sticky ends in recent years. Mike Wansey's WD828 *Mickey's Mouse* was reflown in September 1984 after a lengthy restoration, only to suffer a forced landing near Camden NSW in 1987, sustaining sufficient damage to end its flying days. After restoration to static standards, the aircraft now sits on a pole at the Returned Services League club at Griffith NSW.

This came about after a 'swap' for the previous polesitter, WB518, was arranged, and as noted above, this Firefly at time of writing was undergoing restoration to airworthiness by Classic Aviation at Bankstown.

The other two RAN Fireflies lost in accidents have been WD901 and WH632 and both were owned by the Canadian Warplane Heritage at the time. WD901/CF-BDH flew out of Hamilton Ontario between 1971 and 1977 before crashing into Lake Ontario in a fatal accident. WH632/C-GBDG first flew in April 1992 after a 14 year restoration but crashed after takeoff from Victoria, British Columbia just one month later.

The Sea Fury and Firefly were both significant aircraft in Australia's military aviation history. They were the original equipment of the Fleet Air Arm and both performed admirably on active service. The foundations of the skills and traditions of the RAN Fleet Air Arm rest with them.

Nice shot of Sea Fury WZ647 during an unassisted takeoff from HMAS Sydney. *This aircraft survives at the Camden Museum of Aviation. (RAN)*

Fireflies which fly (or flew) on. Mike Wansey's WD828 'Mickey's Mouse' (top) unfortunately came to grief in 1987, while the RAN Historic Flight's WD826 remains airworthy.

(left) The former WH588 in civilian colours (red and white) as VH-BOU after being sold to Fawcett Aviation in Sydney in 1963. This aircraft now flies with Lloyd Hamilton in the USA. (via Mike Kerr)

Two Former Iraqi Furies which found their way to Australia in the early 1980s: Ted Allen's 'Magnificent Obsession' VH-HFA (top) which now resides in the USA and Guido Zuccoli's '308'/VH-HFG, shown here with its extended canopy over a second seat. Zuccoli imported three of these Furies, selling two and keeping one for himself. (Les Bushell/Stewart Wilson)

(right) The Naval Air Museum's Sea Fury VW623 during its restoration.

Firefly 6 WB518 has been rescued from this pole at the RSL Club at Griffith NSW and in 1993 was undergoing restoration to airworthiness using parts from the damaged WD828. (David Foote)

FLYING the FIREFLY and SEA FURY

These flying impressions and memories of the Firefly and Sea Fury are provided by Commodore Norman Lee, who retired from the Royal Australian Navy in 1981 after a 33 year career. During that time he flew numerous types of aircraft ranging from Tiger Moths to jet fighters and most things in between. He served in Korea (flying Fireflies) and his last posting was as Commanding Officer of Naval Air Station Nowra.

I have fond memories of the Firefly, having flown it for some five years in Fleet Air Arm squadron service back in the 1950s. Being in the right places at the right time I was lucky enough to fly all five operational marks of the aircraft plus the two pilot trainer versions, which included the unique Mk.5 Trainer developed in Australia.

Although now a museum piece, the Firefly in its day was a solid, dependable workhorse which gave sterling service to a number of navies over something like a fifteen year period.

The Mk.I Firefly saw service in the latter stages of WWII and was a follow on from the Fulmar which in turn had followed the Battle. Fortunately, by the time Fairy Aviation got round to the Firefly it had a pretty good idea of what was required in a naval aircraft and the Mk.I and its successors proved to be amongst other things, good deck landing aircraft.

The Firefly was to be in turn a long range fleet fighter, night fighter, strike, and anti submarine aircraft and was well armed for its various roles with four 20mm cannon, bombs, rockets, and assorted ASW weapons.

In developing the Firefly specification, the Royal Navy continued with its requirement for a two seat aircraft with a multi role capability. This followed the relative success of the Fulmar in the Mediterranean during the early stages of the Second World War. The second crew member, an observer in naval parlance, was located behind the pilot and was responsible for navigation, operation of the radar in the night fighter role and of course ASW operations.

As was the routine in those days (1950), following completion of flying training with the RAAF at Point Cook, we were sent to the UK to do our OFS (Operational Flying School) course with the Royal Navy at RNAS Lossiemouth in Scotland and my introduction to the Firefly was via the T.2 trainer which had been developed from the Mk.I. The trainer version of the aircraft followed the usual practise in those days of adding a second pilot cockpit behind the standard front cockpit which in the case of the Firefly was relatively easy because of the already existing rear seat. Unfortunately however, the main fuel tank intruded between the two which meant that the instructor's seat was mounted a fair way aft.

I well recall my first ride in the T.2 if only because I did my best to spin us in. Spinning was a requirement before solo in the Firefly and on the first spin I applied the technique we had been taught in the Wirraway of opposite rudder and then forward stick, with a slight pause in between. As I recall, we were taught rudder before forward stick to prevent blanking of the former by the latter.

The result of my efforts was that the aircraft smartly flicked the other way and I am certain that we would have continued on down flicking from side to side if the instructor hadn't grabbed it off me and rammed the stick into the instrument panel. He must have been watching me like a hawk as a trainer had been spun in on the previous course. I certainly got the message that in the case of the Firefly, it was full opposite rudder and full forward stick at the same time.

Not knowing any better I thought the Mk.I Firefly was a pretty hot ship. It was powered by a Rolls-Royce Griffon 12 and comfortably cruised at 180 knots indicated. As I was subsequently to learn, it was the lightest and fastest of all the various marks of Firefly and was quite good for aerobatics provided you had a strong right arm as the ailerons were particularly heavy.

Our OFS course consisted of the usual formation flying, battle drill, night flying and cross country work. There was a strong emphasis on operating over water and we were soon introduced to sea navigation (called navex's) and carried out quite complicated ship shadowing exercises.

Following completion of OFS part 1 we were posted to RNAS Eglinton near Londonderry in Northern Ireland for weapon

The Firefly T.5 dual control trainer, four conversions of which were performed in Australia.

Commodore Norman Lee renews his acquaintance with the Fairey Firefly in the 1980s after some three decades. The aircraft is the RAN Historic Flight's Firefly AS.6, WD826. (RAN)

training and deck landing qualification. The aircraft used by our next training squadron were Mk.4s which had enjoyed a brief operational period with front line squadrons.

To the casual observer the Mk.4 looked like a Mk.5 or 6 but it didn't have a multi role capability. The most obvious external difference was the lower canopy line which limited head room.

Shortly after we commenced weapon training I had the misfortune to experience the unpredictable nature of early rocket electrical circuits. I was on my way to the firing range and, contrary to instructions, shortly before entering the range I made the armament master switch. Immediately I did so a rocket fired, narrowly missing RAF Ballykelly which I was overflying at the time and the local railway line which ran through the airfield (believe it or not, there were traffic lights at the end of one of the runways!).

I immediately reported the loss and was instructed to return to Eglinton and to carry out a right hand circuit to avoid pointing the remaining rockets at Southern Ireland which was to the north of RNAS Eglinton!

Needless to say, my protestations that I had not touched the firing button were not believed and it was touch and go for a while whether I would remain on course. Fortunately my skin was saved when the aircraft repeated the trick with a senior instructor several days later.

Finally it came time to commence training for the deck.

As most aviation buffs will be aware, in the old straight deck days it was not possible to see the deck when lined up on short finals because of the attitude of the aircraft and the large piston engine blocking the view. This meant that the aircraft had to be "batted" on by the batsman giving signals for speed, height, turn rate and when to cut the throttle. This was done by him using "bats" or "paddles" which looked like oversize ping pong bats.

The deck landing circuit was flown at 400 feet so that the aircraft was at the right speed and hence attitude at the end of the downwind leg. From this point a continuous turn was commenced, gradually losing height with the aim of arriving just short of the stern of the carrier, with the batsman still in full view. This was the tricky bit because if the turn-in was a fraction late, a trailing approach was the inevitable result, with the batsman disappearing under the nose; a very dangerous situation. The problem of when to turn in was exacerbated by the fact that the carrier was steaming into wind and making anything up to 20kts so that when it looked right to turn in, it was too late.

The signal you were looking for as you picked up the bats at the 90 degree point was the "roger" given by the bats being held out horizontally. This meant that you were right for height, speed and rate of turn. Too high was indicated by the bats being held up in a "V, too low by an inverted "V". Fast was indicated by the right hand bat being held out horizontally, slow by the two bats being clapped together.

RAN Firefly AS.6 WD914, one of the last Fireflies delivered for Australian service in 1953. (Richard Hourigan)

The next two signals were the "high dip" and the "low dip". The former, given by the bats being raised into a "V" and then back to the "roger", meant that you were OK but a little too high and required you to momentarily dip the nose down and then to recover to the previous attitude. The "low dip" given by the bats being dropped to an inverted "V" and then back to the "roger", indicated that you were OK but a little low and required you to add a tad of power to regain the correct height.

The signal to be avoided was the "come on", given by the bats being slowly clapped together and scooped up at the same time which meant that you were low and slow. There was no doubting this signal which would have the batsman on bended knee if you had really got yourself into a mess and were heading for the "after roundown". The required response was to add power and to gently decrease attitude until the correct height and speed were regained. The batsman's problem in this situation was that it was dangerous to wave you off because of the possibility of a torque stall if the speed was very low.

The final signals given were the "straighten up" and the "cut". The former was given by the bats being flicked to the horizontal if you were still in the turn, and the latter by the right hand bat being brought smartly across the body in a cutting motion.

The "cut" was mandatory and required you to immediately cut the throttle to idle. The batsman used an imaginary "box" (not unlike the glider technique) in which he could safely cut you. If you were a tad high or fast you would get an early cut, or a late cut if you were a little low or slow. The tolerance was only a matter of a few feet and a knot or two.

Immediately on cutting the throttle, the drill was to let the nose drop to the horizontal which gave you a momentary glimpse of the flight deck and enabled you to assess where you were in the "box".

With experience it was possible to instantly assess your height and line up and to play the flare accordingly. The flare, when initiated required the stick to be brought smoothly and fully back so that the aircraft was in the three point attitude as the hook engaged a wire.

A late flare could result in the aircraft arriving on its main wheels and bouncing causing the hook to miss all the wires. Too great a flare, particularly if the speed was a little high, could result in the aircraft hooking a wire well above the deck and then being slammed down as the wire ran out. It could also result in the hook missing all the wires with a "barrier " the inevitable result.

The barriers were steel nets strung across the flight deck opposite the Island which were hydraulically raised and lowered. They were then to catch any errant aircraft to prevent it plunging into other aircraft parked forward. As soon as an aircraft engaged a wire the barriers would be lowered to enable it to be taxied into the forward deck park.

Our training for the deck consisted of hours and hours of aerodrome dummy deck landings or ADDLS under the control of the training batsman. We had to learn to confidently fly our aircraft by following his signals, and he had to be satisfied that we could do this precisely and consistently.

Speed for the deck landing approach in the Firefly was 90 knots and from memory -2lbs of boost was about right. Even though we were under the control of the batsman, I developed a scan that included the ASI and the boost gauge and found that the best throttle control was provided with my forearm resting on the canopy tracks and protruding out in the breeze, working the throttle with the tips of my fingers.

Some aircraft had decklanding ASIs fitted in a bulge outside the cockpit and in your line of sight to the batsman. They only read about ten knots (from memory) either side of the required speed. I never found

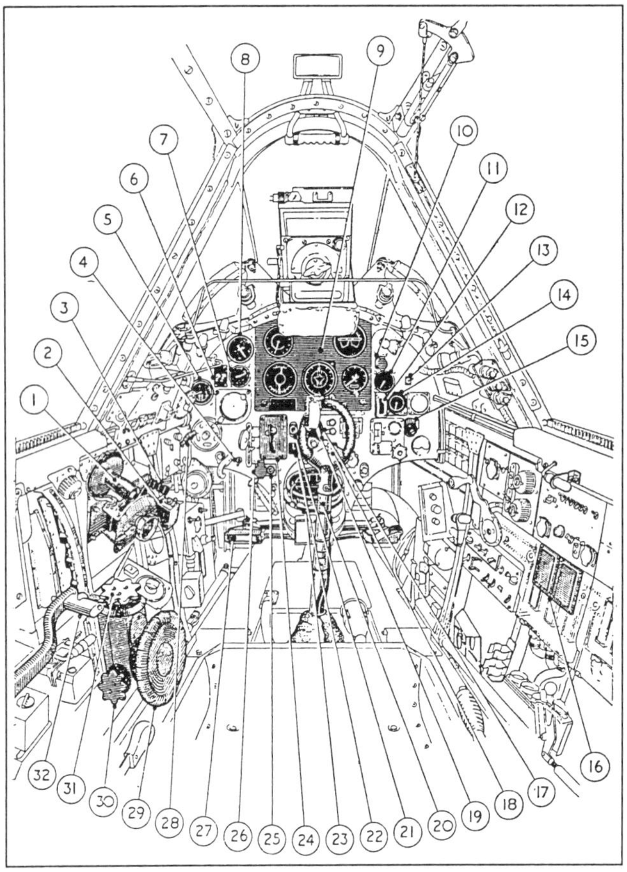

FIG. 1 FLYING AND ENGINE CONTROLS: 1 – throttle; 2 – supercharger selector switch; 3 – propeller speed; 4 – air cleaner; 5 – brake pressure triple gauge; 6 – ignition switches; 7 – boost gauge; 8 – tachometer; 9 – instrument flying panel; 10 – oil dilution push button; 11 – coolant temperature gauge; 12 – oil pressure gauge; 13 – priming switch; 14 – oil temperature gauge; 15 – starter magazine reload; 16 – engine data plates; 17 – fuel contents gauge; 18 – air intake switch; 19 – brake lever; 20 – control column; 21 – fuel pressure warning lamp; 22 – fuel pump booster switch; 23 – P.4A compass; 24 – fuel cock; 25 – slow running cutoff; 26 – starter push switch; 27 – rudder bar; 28 – radiator shutter selector switch; 29 – elevator trim control wheel; 30 – aileron trim control wheel; 31 – rudder trim control wheel; 32 – flap lever.

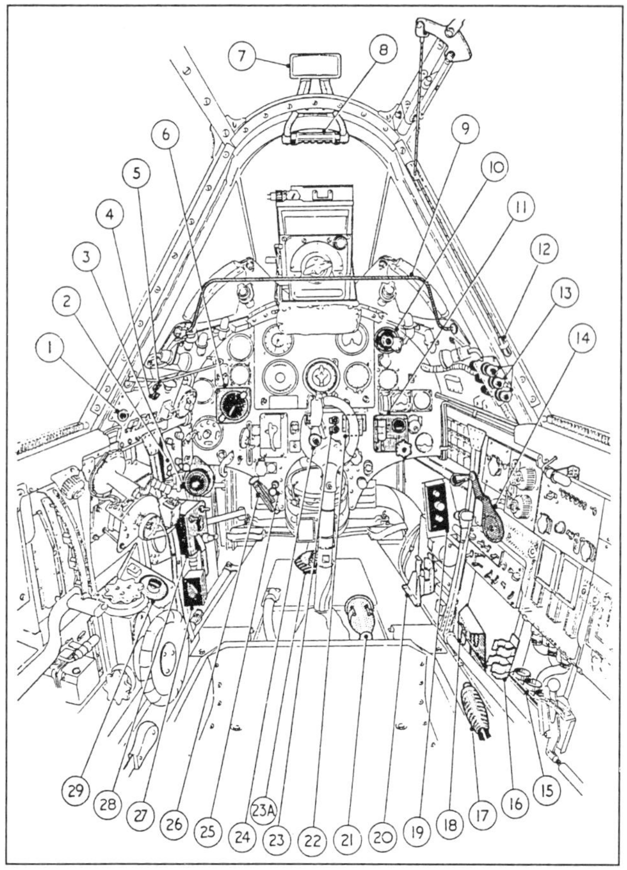

FIG.2 MISCELLANEOUS CONTROLS and EQUIPMENT: 1 – Observer call push button; 2 – ARI.5307 control; 3 – Mk.3A altimeter contacting; 4 – telephone warning switch; 5 – RT muting switch; 6 – ARI.5284 altitude indicator; 7 – rear view mirror; 8 – ventilator; 9 – chartboard support (stowed); 10 – windscreen deicing control; 11 – oxygen regulator; 12 – canopy jettison handle; 13 – ARI.5284 altitude limit indicator; 14 – canopy operating handle; 15 – signal cartridge stowage; 16 – torch stowage; 17 – seat adjusting lever; 18 – camera stowage; 19 – Type 383 control; 20 – signal pistol stowage; 21 – 'relief' tube; 22 – generator field circuit breakers; 23 – power failure warning lamps; 23A – foot operated ventilator; 24 – press to transmit switch; 25 – heating control; 26 – pilot's notes stowage; 27 – ARI.5307 junction box; 28 – ARI.5669 control; 29 – ARI.5284 altitude limit switch.

them to be of much use because they were rare and required an altered scan.

After some weeks of ADDLS we were judged as being ready for the deck and duly joined HMS *Illustrious* (the earlier one) in the Irish Sea for deck landing training or DLT as it was known.

I'm afraid my first landing was rather heavy due to an incomplete flare resulting in a bounce and the airborne runout of the wire. This resulted in some damage to the torsion box, the only time I bent an aircraft for the rest of my deck landing career.

Not so lucky was my mate "Blue" who, on his first landing, failed to let the nose drop on receiving the cut and immediately flared. The result was that he missed all the wires and ended up going through the barriers at some speed, shedding his undercarriage, flaps, airscrew and other bits and pieces, stopping just short of the forward 4 inch guns which were inset in the flight deck in the Illustrious class. It was the first barrier I had witnessed and was remarkable for the sight and very expensive sound.

With several sessions of DLT under my belt, I was just getting to think that I knew what I was doing when I had my confidence rudely shaken. On picking up the bats I saw that I was being given a slow. I added power but still continued to receive a slow and shortly afterwards a "come on". At this point I had the throttle way forward of where it should have been but even so, I could feel the aircraft settling.

As the adrenalin started to surge I had the good sense (training?) to quickly check the cockpit and to my relief found that the pitch lever had slipped back. This was a problem with early Fireflies where if the throttle friction nut was adjusted for smooth operation of the throttle, which was essential for deck landing, it was insufficient to hold the pitch lever. At this point I was very low and slow and very conscious of the dreaded torque stall should power be applied too quickly, but at the same time daren't bring the throttle back before advancing the pitch lever as I needed what little power I had. All I could do was advance the pitch lever as smoothly as possible until normal power was restored. If I had rammed it forward there is little doubt that the aircraft would have flicked onto its back and I would now be in the middle of the Irish Sea.

Our first takeoffs from the deck during our DLT were free takeoffs as opposed to being shot off via the catapult. Even though we had the whole deck to ourselves, it was still a bit daunting as the forward end of the flight deck looked awfully close. An added mental problem was the "Island" which was only a matter of feet away at the closest point on your run down the deck.

The Rolls-Royce Griffon 74B in the later marks of Firefly developed 18lbs (about 88ins) of boost at full chat and the torque was considerable. For a free takeoff, the aircraft was held on the brakes and the throttle advanced to a maximum of about

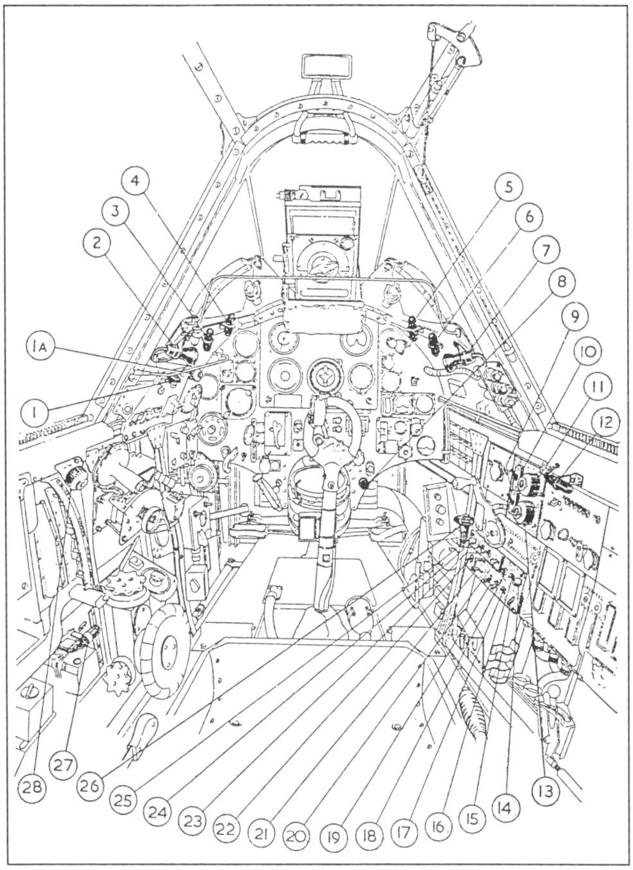

(left) FIG.3 LIGHTING: 1 – port switch panel flood lamp; 1A – dimmer for external ASI; 2 – instrument ultraviolet lamp (port); 3 – instrument flood lamp (port); 4 – emergency instrument lamp (port); 5 – emergency instrument lamp (starboard); 6 – instrument flood lamp (starboard); 7 – instrument ultraviolet lamp (starboard); 8 – compass flood lamp dimmer switch; 9 – emergency lamps switch; 10 – cockpit flood lamps master switch; 11 – instrument ultraviolet lamps dimmer switch; 12 – armament panel flood lamp; 13 – instrument flood lamp dimmer switch; 14 – Morse key; 15 – recognition lamps switch; 16 – upward identification lamps switch; 17 – recognition lamps switch; 18 – resin lamps switch; 19 – navigation lamps switch; 20 – formation lamps dimmer switch; 21 – navigation lamps dimmer switch; 22 – navigation lamps switch; 23 and 24 – formation lamps switches; 25 – Morse key; 26 – lighting panel flood lamp; 27 – trim control box flood lamp; 28 – cockpit flood lamp dimmer switch.

9lbs (depending on the deck space available). At this point the tailwheel would be bouncing on the deck and there was no doubt that the aircraft wanted to go. On release of the brakes it was essential to really stand on the left rudder pedal to prevent a swing to the right developing. If you lost it, it was not possible to recover from the swing without reducing power which was the last thing you wanted to do. You also had to exercise a little care in getting the tail up because the gyroscopic effect exacerbated the swing. Provided you had everything under control, the rest of the takeoff was fairly straight forward and the aircraft would fly off with plenty of room to spare.

I only had one worrying free takeoff which took place a year or so later when the squadron was flying off to NAS Nowra. My aircraft had gone unserviceable and I was detailed to fly the spare ashore without my observer. Suddenly I had lots of friends as the Air Group discovered that I had an empty rear seat. I ended up with more golf clubs than Greg Norman plus other assorted items to the point that it was difficult to close the rear hatch. Through all this I must confess that I never gave the centre of gravity a thought.

To make matters worse, my aircraft was ranged right at the front of the deck park as was usually the case for the junior member of the squadron. As I manned the aircraft, the forward end of the flight deck looked very close.

The end result was that I literally fell off the front end with the aircraft well tail down and with forward trim going on as fast as I could wind it. The dependable old Firefly gradually clawed its way into sustainable flight with me having learnt another lesson about flying, and more to the point, survived it.

After a few free takeoff sessions we were introduced to the catapult which in those days was still hydraulically operated; the steam catapult which gave you a smooth progressive ride was still some years away. Being hydraulic, the power of the shot was applied almost instantly, not unlike being kicked up the behind by an elephant.

The drill was not to hold the stick, but to keep your hand braced behind it so that you didn't pull it back as the acceleration took effect. Despite all the briefings, on my first shot I wasn't ready for the rapid ride and my arm slipped back. Instinctively I pushed forward but unfortunately over reacted, with the result that the aircraft went off the catapult heading for the water. Having disappeared below flight deck level, the resultant recovery of a Firefly in almost plan view was evidently very spectacular!

Following completion of our OFS we returned to Australia to join 817 Squadron of the 21st Carrier Air Group which was operating Mk.6 Fireflies in the purely ASW role. The Mk.6 had no cannon which was a great disappointment for a gungho young sub lieutenant; there is nothing like the rattle of four 20mm cannon to stir the blood.

The aircraft was capable enough in the ASW role but was limited by the ASW sensors then available which consisted of the Mark One eyeball, and non directional sonobuoys.

Fortunately, shortly after I joined the squadron it was nominated to be part of the Sydney Air Group for operations in Korea and we immediately swapped our Mk.6s for our sister squadron's Mk.5s and commenced an armament workup. Our role was to be bombing and straffing whilst the two Sea Fury squadrons rocketed and also strafed.

It was decided that we would dive bomb as our main targets would be bridges. The attack profile consisted of a 55 degree dive commencing from 8,000ft with release of the two 500lb bombs at 3,000ft. The Firefly was never really meant to be a dive bomber and picked up speed very quickly in the dive. If you didn't get the trim right at the top of the dive it was almost impossible to hold it on the correct flight path, often times it felt as if though we were going down sideways because of the large skid angle.

As it turned out, when we commenced operations, even though we were accurate enough to achieve a straddle, the bridges would remain standing. A punt was then taken that we would get away with a low level attack, aiming by eye and diving diagonally across the bridge to put the bombs into the abutment. As we were attacking in flights of four, it was necessary to use delay fuses so that the last aircraft could safely bomb. With only a 35 second delay it was no time to be tardy if you were tail end charlie, which was normally my position in the flight.

With experience we became quite good at the job and could knock a bridge down with only one aircraft. The bridges were relatively flimsy affairs and would be re-erected by the North Koreans overnight. The next day we would do our rounds and knock them all down again!

We only lost one aircraft in this form of attack flown by my mate "Blue" which was shot down and resulted in the longest and most dramatic rescue of the war.

At that stage of the war there was very little surface movement in our area of interdiction so straffing targets were few and far between and one had to be wary that a juicy target wasn't a flak trap, as one senior squadron member found to his embarrassment.

With the cynicism and naivety of youth, I must admit that I looked upon our time in Korea as a sort of finishing school to our OFS.

Several years later I was posted to the UK for Gannet conversion and squadron experience with the RN. During this period an opportunity arose to do some observer training in the Mk.7 three seater Firefly. This aircraft was fitted with a Rolls-Royce Griffon 57, the same engine fitted in the Shackelton but without a contra rotating prop.

The aircraft was quite different in appearance to previous marks of Firefly having a large rear cockpit and very tall fin, nevertheless it handled much the same and was pleasant to fly.

Throughout almost a 1,000 hours in Fireflies, I never experienced any serious problems with the Rolls-Royce Griffon in its various marks. I did have one stop as I retarded the throttle after takeoff but it ran again as soon as I advanced the lever to the fully open position. The problem turned out to be some foreign matter in the injector and was not the fault of the engine.

We had an amusing although frustrating period of operations when RR discovered that a batch of cams hadn't been properly hardened and had ended up round in a very short period of time. An STI (special technical instruction) was issued requiring the cams to be checked every five hours until each engine received a clean bill of health. I can't recall my squadron having any of the faulty cams.

Engine start was achieved through the time honoured naval system of the Coffman cartridge starter, which once you had established how much to prime, and the correct setting for the throttle, you normally achieved a start first time.

As I said at the beginning, the Firefly was a solid, dependable aircraft which handled well in the air and on the ground. Its large Fowler flaps, controlled by a very large lever on the left hand side of the cockpit, added considerably to the wing area in the three positions of takeoff, cruise and land. We never used the cruise position, which had the flaps out, but at the same angle of attack as the wing, as it didn't improve the performance in any way.

The cockpit was roomy enough and the view was good as you sat well forward with the bulk of the wing behind you. Trim was provided on the three axis by three wheels just under your left hand and it was possible to set the aircraft up for short periods of hands off flight. When deck landing, the drill was full left rudder trim and I found that a little left aileron helped with some aircraft.

Mixture control was automatic, and being a British aircraft, the boost set would be held to the rated altitude through the ABC (automatic boost control). The two speed supercharger could be set to change to high gear automatically as you climbed to height, or you could change it yourself, reducing power as you changed to save the engine receiving a solid thump.

Wing fold in the early Fireflies was entirely manual and required the deck team to pull a large lever out of the leading edge and rotate it downwards to withdraw the locks. Large poles were then hooked into recesses in the wingtips and the wings were then folded back by teams on either side. With the later marks of aircraft, the same procedure was required to withdraw the locks but the wings were then folded hydraulically.

It was always comforting flying the Firefly knowing that there was no way that the wings could fold in flight. All other naval aircraft relied on a manual lock on the wing fold selector and the usual squat switch on the undercarriage. I must confess that on my first ride in a Sea Fury I was a little apprehensive that I had got the wing spreading procedure right!

The Firefly was easy to control on the ground although excessive use of the brakes was to be avoided to prevent heat build up in the air bags of the pneumatic braking system. The large rudder enabled the aircraft to be manoeuvred using short bursts of power but care had to be exercised that speed didn't build up.

Well that just about sums up the Firefly, as I said at the beginning I have very fond memories of the aircraft in which I served my apprenticeship as a naval aviator. Let's hope that the Fleet Air Arm Museum manages to keep its remaining Firefly, WD 826, airborne for many a long year to come.

What happens if the wires are missed: "remarkable for the sight and very expensive sound ..." A Firefly pilot gets it wrong aboard HMAS Sydney.

RAN Sea Fury WH587 poses for the camera.

Sea Fury

After a couple of years flying Fireflies, including a tour in Korea, I managed to wangle a conversion to the Sea Fury. It was the first of three of Sir Sydney Camm's designs that I was lucky enough to fly. The other two were the Seahawk and the Hunter.

After digesting the pilot's notes and a thorough briefing on how to handle the Bristol Centaurus 18B engine, an entirely different technique to the Griffon, I was sent on my way.

The Centaurus was fitted with an interlinked throttle and pitch control system which automatically selected the correct RPM to match the boost selected. This was a great asset in a fighter where rapid power changes are required in combat. The only snag for someone who was conditioned to the principle of "throttle before pitch" when reducing power, was that before the throttle was retarded after takeoff, the pitch lever had to be moved to the full coarse pitch position to bring the matched boost/ RPM system into operation. No drop in RPM occurred until the throttle was reduced.

Like the Griffin in the Firefly, the Centaurus was fitted with a Coffman cartridge starter which required the engine to be carefully primed and the throttle fairly precisely set if you were to get a start first go. The Coffman starter was fitted with a safety disk which would blow if too much pressure built up immediately after the cartridge was fired. The system held six cartridges which were about the size of a shotgun cartridge but slightly longer. All good naval aviators in those days carried a spare safety disk in case the disk blew on a land-away somewhere, a not unknown occurrence (a shilling would do but was not recommended by the manufacturer!).

As I recall my first few flights in the Sea Fury, what impressed me most was the acceleration and the lightness of the controls. The ailerons were remarkably light after the Firefly due to the spring tab system

Fuelling up a Sea Fury before flight. WH588 was one of the aircraft painted in a US Navy style dark gloss blue for display purposes. (RAN)

(right) *One that missed the wires aboard HMAS Sydney (via Keith Western)*

adopted by Hawker. With this system the control column was not directly connected to the ailerons but to spring tabs on each aileron which when deflected, in turn deflected the aileron itself. A similar system was to be fitted to the Gannet.

After the Firefly I always felt slightly hemmed in in the Sea Fury as the cockpit sides were level with your shoulders with only your head protruding into the canopy. In the good old Firefly you could rest your forearm on the canopy rails, a great asset in deck landing as it gave you more precise throttle control.

Another major difference in the two aircraft was that you sat almost level with the trailing edge of the wing in the Fury whereas in the Firefly you were well forward.

For sheer joy, there was nothing better than to haul the beast around in a steep turn, the thirteen feet of nose in front of you acting like a great big pointer. The trick was to place a certain cowl fastener on the horizon and you were guaranteed to hit your slipstream every time.

The greatest handling difference between the two aircraft was in the aerodrome landing technique. With the Firefly, there was sufficient elevator control to play the flare to achieve a smooth arrival; the aircraft also sat close to the landing attitude at 90 knots. With the Sea Fury, you only got one go once you had cut the throttle as there was insufficient elevator to hold that great nose up for more than a second. It was a case of ensuring that the speed was right on 100 knots and then fairly smartly back with the stick as you cut.

We Firefly drivers found the rudder somewhat sensitive on landing and in my first few rides I confess to weaving down the runway as I overcontrolled during the final stages of the landing rollout. As in the Firefly, the Fury was fitted with hand operated brakes which were fairly powerful for their day and to try and stop the weave with brake normally only made the problem worse.

The most important instrument in the Sea Fury was the oil pressure gauge, anyone who has flown a sleeve valve engine will know what I mean. Failure of the oil supply could result in rapid engine seizure and with that big five bladed prop, a forced landing could be tricky indeed.

It was a rule when on a session of touch-and-goes to do a cooling circuit if the cylinder head temperature was up at all, particularly in hot weather. Early in my conversion to the aircraft, during a session of circuits, I noticed the temperatures were up a bit as I turned downwind but thought I would be OK. Sure enough as I was cleaning up following the touch-and-go I saw that the oil pressure had dropped to 60 PSI; anything under 100 was cause for concern. As soon as possible I throttled back trying to gain maximum cooling effect with minimum power, promising the Diety above that if He would overlook this one, I would never do it again. To my relief, the cylinder head temperature dropped and the oil pressure returned to normal. I landed and put the aircraft unserviceable, but lucky for me the engine checked out OK.

Mention of snagging the aircraft after fight reminds me of that famous entry in the Form 700, "Engine sounds like the clappers of Hell". The clearing entry read "Clappers removed, engine serviceable".

I must admit to giving myself quite a scare in a Sea Fury on one occasion as a result of overconfidence coupled with poor judgement. Shortly after we had returned from Korea, I had been sent off to sea to gain a watch-keeping certificate in HMAS *Murchison*, a River class frigate. Half way through my sea time I requested a period of refresher flying which was granted.

Murchison then sailed for Jervis Bay and I had the pleasure of joining her in a successful anti-submarine exercise, flying a Firefly. The next day I managed to scrounge a Sea Fury for some GFP and couldn't resist finding the ship and giving the crew a demonstration of my skills.

After a few fast runs I decided to do a low, slow fly-by with everything down, but failed to take account of the ship's speed and the strong wind into which she was steaming. I had intended to fly across her bows but as I closed in, I found that I was having to increase my bank angle in order to pass ahead. Things got rapidly worse and soon I had so much bank on that I lost sight of her. It was one of the fastest clean-ups I've ever done and with full power on, I roared across her fo'csle with a few feet to spare and climbed away a very chastened young man.

I've often pondered on how luck must play a part in every aviator's development, a few seconds more and I would have written myself off. However, I survived and was a much wiser pilot as a result. Years later at *Murchison* reunions I would be told how I had impressed everyone onboard; I would mumble something and quickly change the subject!

Finally my one regret was that I never had the opportunity to fly the aircraft operationally, but from my limited experience I can say that without doubt, the Sea Fury was one of the finest piston engined fighters ever built.

THE FLEET AIR ARM IN KOREA

The division of Korea into North (communist) and South (democratic) elements came about as a result of the end of World War II and the splitting of the previously Japanese held country into American and Soviet controlled halves.

Attempts to unite the two halves in the late 1940s and early 1950s had failed, each claiming to be the one true and legal government. Accordingly, tensions rose, and the communist North indulged in subversive activities coupled with a programme of propaganda designed to undermine the will of the South Koreans to remain associated with the United States and the rest of the Free World.

Subversion turned into full military action in the early morning of 25 June 1950 when the North Korean Peoples' Army (NKPA) launched attacks across the border between the two Koreas. A breach of the peace was declared by the United Nations and member countries were called upon for assistance.

Australia's immediate response was to assign the frigate HMAS *Shoalhaven* (based in Japan) and the destroyer HMAS *Bataan* (in Hong Kong) to the allied forces along with the RAAF's No 77 Squadron which was based in Japan as part of the British Commonwealth Occupation Forces there, initially flying Mustangs and then Meteors.

Other Australian warships were directly involved in the Korean 'Police Action' (as it was sometimes described) as were large numbers of Australian soldiers and airmen. The conflict ended in July 1953 with the two Koreas remaining separated as before, although at least the feared wider conflict – which seemed inevitable when China became directly involved – was avoided.

The Royal Australian Navy lost three men during the Korean war, all of them airmen from the aircraft carrier HMAS *Sydney*.

Carrier to Korea

On 11 May 1951, the Australian Prime Minister, Robert Menzies, announced that the aircraft carrier HMAS *Sydney* and her Air Group comprising two Sea Fury (Nos 805 and 808) and one Firefly (No 817) squadrons would sail to Korea to relieve the Royal Navy's HMS *Glory* for a period of about three months from October 1951. *Glory* was one of several British light fleet carriers involved in the Korean war and would sail to Australia for an extensive refit at Sydney's Garden Island dockyard while HMAS *Sydney* took her place.

817 Squadron was equipped with Firefly Mk.5s for its duties in Korea, relinquishing the Mk.6s with which it had previously been issued as the latter did not feature built in cannon armament, a necessity for the types of operations about to be undertaken.

HMAS *Sydney* departed her namesake port on 31 August 1951 under the command of Captain David Harries. The commanding officers of the three squadrons of aircraft were Lt Cdr W G Bowles (805), Lt Cdr J L Appleby RN (808) and Lt Cdr R B Lunberg RN (817). The Sydney Air Group commanding officer was Lt Cdr Michael Fell.

Sydney travelled to Korea via Manus Island and Japan, refuelling taking place at the former along with practice army support and spotting support sorties by the Sea Furies and Fireflies. Earlier in the voyage, the carrier had entered (but not anchored in) Rabaul Harbour in order to 'show the flag' and put up several formations of aircraft.

Japan was reached on 19 September, the ship stopping at Yokosuka for eight days before sailing on to Kure on southwest Honshu. There, the handover from HMS *Glory* was carried out, an involved affair requiring the transfer of several tonnes of maps, briefing data, aerial photographs, escape and evasion notes and so on. The attention to detail of the handover assured *Sydney* of a smooth transition to operational duties, as did the thorough working up period beforehand.

HMAS *Sydney* departed Kure on 3 October, almost ready to begin operations just two days later under the auspices of Operation 'Strangle', which was designed to interrupt the enemy's supply lines by striking at his means of transport. That first

Sea Fury VW636 catches the wire on board HMAS Sydney. *The pilot was Sub Lt Ronald Coleman, who was killed in action on 2 January 1952 when his Sea Fury went missing during a CAP mission in Korea. (via Richard Hourigan).*

patrol would be conducted off Korea's west coast, and all operations would be conducted as part of Task Group 95.1. *Sydney* would alternate with the US light carriers *Rendova* and *Bodoeng Strait* during her seven patrols, most of which were conducted off Korea's west coast in the Yellow Sea.

In the meantime, all aircrew attended lectures on escape and evasion, dummy briefings were carried out and numerous meetings and conferences were held along with a high degree of planning for the first days' operations. A sidelight for the Firefly crews was instruction in the use of the Owen sub-machine gun, issued in case they were forced down over enemy territory.

Summary of Operations

The day to day operations of the Australian Sea Fury and Firefly squadrons in Korea are described in more detail below, it is perhaps appropriate to present a summary of those operations first.

Between 5 October 1951 and 25 January 1952 the aircraft flew 2,366 sorties for the loss of eight Sea Furies and three Fireflies with three aircrew killed. The Fireflies of 817 Squadron contributed 743 of the sorties (including 179 anti submarine patrols), while the combined efforts of 805 and 808 Squadrons' Sea Furies resulted in 1,623 sorties. The average daily sortie rate for aircraft was 52.2.

This translated into a daily sortie rate of about 1.5 per pilot, although taking into account pilot unavailability, this increased to a typical two sorties per day per pilot. Most pilots averaged about 30 hours flying per month during Korean operations, although this could vary considerably. The high sorties rate was possible not only because of the ship's good state of readiness, but also because of a complete lack of enemy aircraft and naval opposition to be concerned with.

The ship's standard patrol period was ten days with one 'day off' in the middle for refuelling by one of the British Royal Fleet Auxiliary ships operating in the theatre.

Attrition replacement aircraft came from the British carrier HMS *Unicorn*, which acted as a maintenance ferry carrier, bringing replacement aircraft and stores from Singapore to Japan and thence to the operational ships. Replacement aircraft were taken on board *Sydney* (and the other carriers) when she was in port.

Unicorn's 'double decker' hangar design made her ideal for this role, although during the conflict she was occasionally used to accompany the other aircraft carriers into the combat area, flying her own aircraft and providing an extra flight deck.

The squadrons' standard aircraft strength was 12 each plus two reserve. Between them, the three squadrons' aircraft dropped 410,000lb (186,000kg) of bombs and fired 6,359 rockets and 270,000 rounds of 20mm ammunition. Climatic conditions ranged from very hot and humid to arctic with temperatures well below zero. An extremely violent typhoon also had to be contended with.

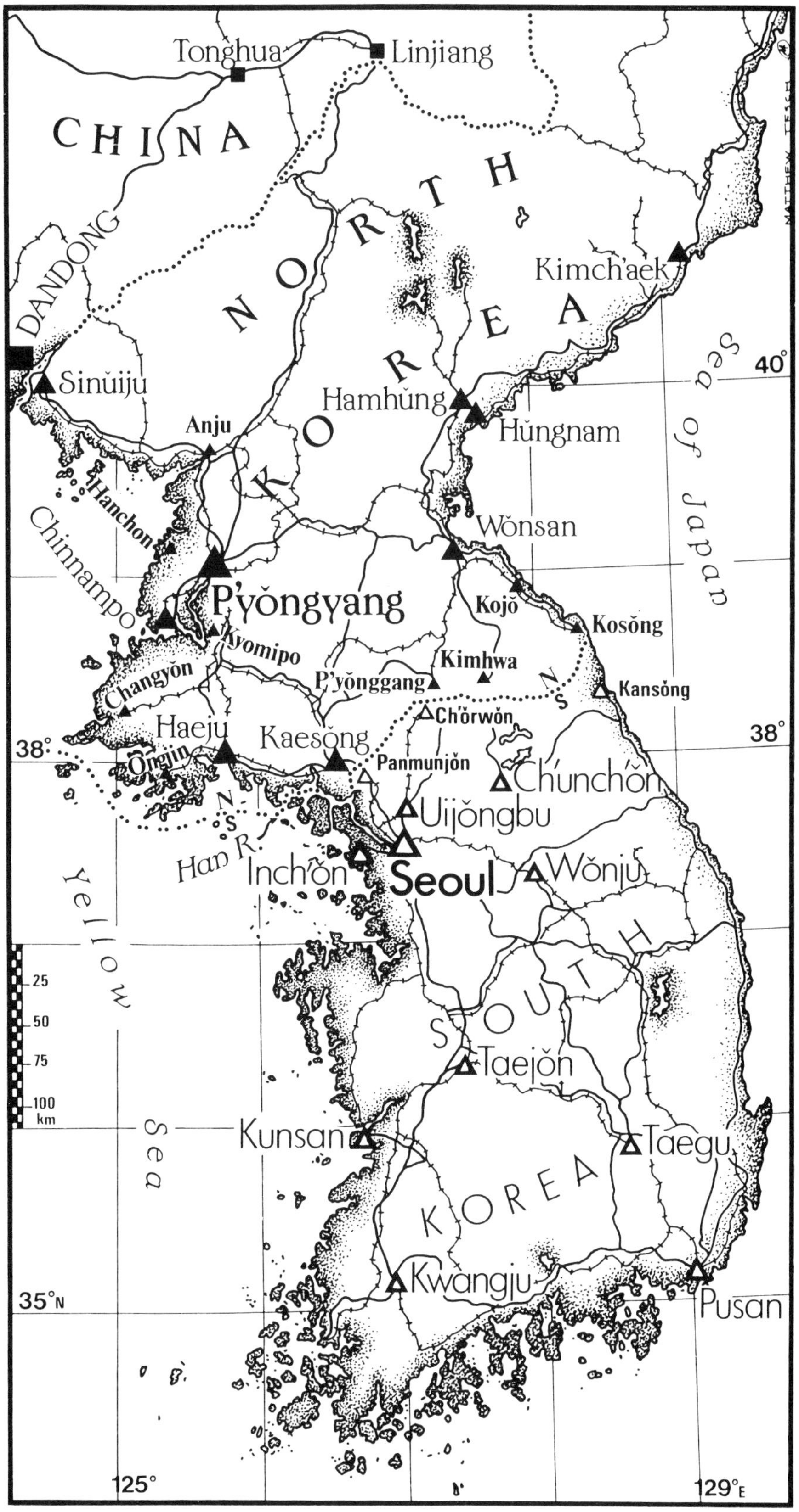

The 'K' on the tails of these Sea Furies indicates they operate from HMAS Sydney, on the water below, but the aircraft in the foreground, VW543, carries Royal Navy titles. Several Sea Furies (and Fireflies) which were loaned to the RAN from the Royal Navy flew with the Australian squadrons during the Korean war and were never officially taken on RAN charge. (RAN)

Busy scenes on the flightdeck of HMAS Sydney as rocket armed Fireflies prepare to launch.(RAN)

The Australians inflicted considerable damage on the enemy, as the following figures testify:

	Dest'd	Damaged
Troops	1428	1701
Vehicles	7	8
Field guns	7	7
Rail bridges	47	13
Road bridges	4	4
Rail tunnels	—	7
Rail cuttings	38	—
Rolling stock	21	118
Locomotives	1	1
Water towers	—	5
Buildings	1054	1246
Junks	36	143
Sampans	66	222
Ox carts	234	155
Stores dumps	11	39
Fuel dumps	7	—
Ammunition dumps	16	—
Mortars	13	—
Mobile cranes	—	3

Four RAN Sea Furies in formation, three of them wearing black and white Korean identification stripes. (RAN)

The Sydney Air Group performed many and varied tasks off both coasts of North Korea during its tour of duty including the following:

Close Army Support: The Sea Furies flew many missions in support of the Commonwealth Division (often 3rd Battalion, Royal Australian Regiment) and the USA's 1st Cavalry Division. Targets were usually troop positions close to friendly forces, requiring very accurate flying and gunnery.

Bombardment Air Spotting: Numerous sorties were flown by both the Sea Furies and Fireflies, spotting for ships ranging in size from RAN frigates to US battleships. American procedures were used and the system worked well when the fact that operations in conjunction with American, British, Australian, Canadian, Dutch and New Zealand ships were carried out. Targets ranged from single gun positions to large villages and towns.

It was quickly discovered the smaller ships could not provide fire accurate enough to be effective against pinpoint targets. While the fire from battleships and cruisers was usually accurate, destroyers and frigates were most effective against area targets against which less accuracy was adequate.

Rail Interdiction: The primary role allocated to the Fireflies, using dive bombing (at first) and then low level bombing techniques in combination with 27 second delay fuses.

Guerilla Targets: These were usually troop concentrations in villages and other camouflaged hideouts. Information on the whereabouts of these targets was provided by friendly guerilla forces in the area and the Sea Furies were used against them, strafing with guns and firing rockets. These operations were known as 'Leopards', after the codename given to the guerilla commander on the ground.

Anti Shipping Strikes: A daily reconnaissance of the coastline and main waterways in the operational area was carried out by Sea Furies. The idea was to enforce the blockade of North Korea by attacking enemy junks and sampans. These operations resulted in the thwarting of several communist amphibious assaults against friendly islands.

Armed Reconnaissance: Once the Sea Furies' and Fireflies' primary targets had been attacked, the aircraft patrolled a prearranged area to fly armed reconnaissance and to attack targets of opportunity. The result was that every road and rail line within HMAS *Sydney's* area of responsibility was patrolled on a daily basis.

Members of the Sydney Carrier Air Group pose for a family portrait at the end of Korean operations. Seated in the front centre is Lt Cdr Michael Fell (Air Group Commander), to his right Lt Cdr R B Lunberg (817 Sqdn CO) and to his left Lt Cdr W G Bowles (805 Sqdn CO) and Lt Cdr J L Appleby (808 Sqdn CO). (via Norman Lee)

The night of 14 October 1951 saw HMAS Sydney *hit by Typhoon 'Ruth' off Korea. The result was one Firefly washed overboard and six other aircraft badly damaged plus other damage to the ship and its equipment. This series of photographs were taken during the storm and the morning after. (RAN/Joe Barr/ Keith Western)*

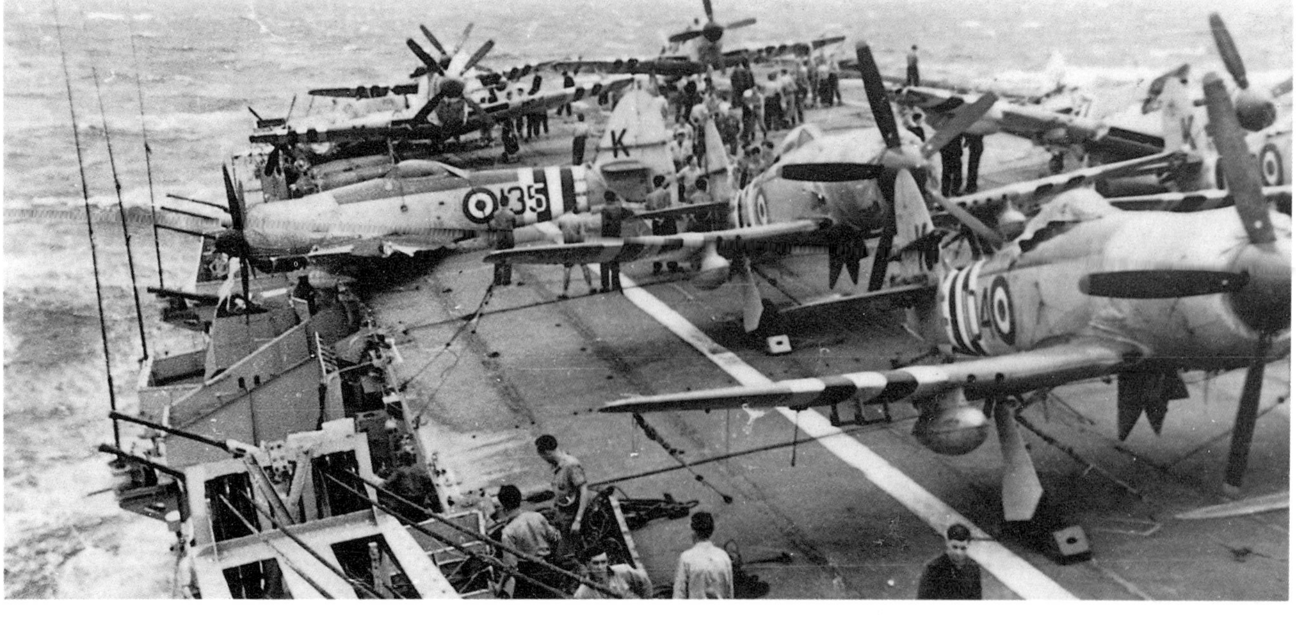

Aerial Photography: Using vertical and oblique cameras, the Sea Furies photographed all the communications centres, rail bridges and major towns in the operational area so as to select targets and also for use in briefing and damage assessment.

Observers in the Fireflies also took some hand held oblique photographs, mainly for damage assessment. Some 5,000 aerial photographs were taken during *Sydney's* period of operational duty.

Defensive Missions: Combat Air Patrol (CAP), Anti Submarine Patrol (ASP or sometimes ASPRO) and Convoy Combat Air Patrol (CONCAP) sorties were flown, but without any contact with the enemy.

Escape and Evasion was an important part of the Fleet Air Arm's Korean excursion. If a pilot was downed, at least one helicopter would be sent to pick him up and other aircraft would provide cover in the form of a Combat Air Patrol (CAP). All pilots who survived being forced down or bailing out over North Korea were successfully recovered.

HMAS *Sydney* carried a US Navy Sikorsky S.51 'Dragonfly' helicopter for these duties, flown by an American pilot.

A Firefly into the barrier on HMAS Sydney. *(RAN)*

Korean War Diary

The following is a daily summary of the Sydney Air Group and its main activities during its tour of duty between early October 1951 and late January 1952. The information is derived from the ship's 'War Diary – Korean Operations', and provides an interesting insight into the day to day activities of the Sea Furies and Fireflies:

5 October 1951: The first day of operations was marred by fog which prevented attacks on the primary target (ferry terminals at Taedong-Gang and a road bridge at Chinnampo) but four Furies and four Fireflies attacked other targets, the former damaging buildings at Ongjin with rockets and the latter attacking a road bridge at Songwha.

In addition, troop shelters and one large building were destroyed and junks, houses and rolling stock damaged. Enemy troops harassing friendly forces were also strafed.

A Bombardment Spotting Shoot was attempted without success due to the inaccuracy of the frigate HMS *Black Swan's* gunnery. Overall, 47 sorties (32 offensive and 15 defensive) were flown, including Combat Air Patrols (CAPs) by Sea Furies and Anti Submarine Patrols (ASPs) by depth charge armed Fireflies.

6 October: 32 offensive and 15 defensive sorties flown with one aircraft suffering light flak damage. Two Furies carried out an Air Spot with HMS *Cossack* in the Haeju area with good results.

A Firefly knocked out a span of a major road bridge at Chinnampo, several buildings were destroyed and others set on fire. The Fireflies' standard offensive load was two 500lb (227kg) bombs and cannon, while the Sea Furies normally carried eight rocket projectiles in addition to their internal guns.

7 October: Only 23 sorties as the ship refuelled in the afternoon. Activities included the bombing of a rail bridge at Haeju, a rocket attack on a barge south of Kyomipo and a successful impromptu air spot with the frigate HMS *Amethyst*. Three Sea Furies were damaged by flak, one of which had to land at Kimpo with no rudder control. Small arms fire from the ground would continue to be a major problem for the aircraft.

On this day, *Sydney* was ordered to proceed to the east coast of Korea for two days' operations in the Kojo area, the objective being to destroy installations and defences, possibly preparatory to amphibious landings.

8-9 October: Ship *en route* Kojo area.

10 October: 55 sorties flown, many of them against targets in the Kojo area with damage estimated as seven large barrack buildings destroyed, 18 troop shelters and six boats damaged and destroyed and the main railway line cut in several places. Shoots were carried out with HMS *Belfast*, HMS *Concord* and the USS *Colahan.*

11 October: A very busy day with 89 sorties flown, a record for a light fleet carrier in the Korean war. Three aircraft damaged by flak. Supply and bivouac areas were attacked and the villages of Chonchon and Tonghow nearly destroyed. Shoots with HMShips *Belfast* and *Concord* plus the USShips *Hansom* and the battleship *New Jersey* continued, the latter's 16 inch guns destroying the villages of Korin-Ni and Singhung-Ni.

During the shoot against Singhung-Ni, the pilots reported seeing about 1,000 enemy troops and stores dumps on ridges in the Kojo area, apparently reinforcements being built up in anticipation of a landing. Both the troops and the dumps were strafed by Sea Furies, with enemy casualties estimated at some 200.

The effort put in by *Sydney's* Sea Furies and Fireflies on this day was recognised bby congratulatory signals, including from the British Chief of Naval Staff: "I would add my congratulations to a personal message I have received from the first Sea Lord conveying his congratulations on HMAS *Sydney's* outstanding performance on her first operational patrol".

And from the Commander-in-Chief: "Your air effort in the last two days unprecedented in quality has been a magnificent achievement on which I warmly congratulate you. Though it is invidious to particularise, the spotters especially did a first class job ... 89 sorties in one day is grand batting by any standards, particularly in the opening match".

For more detailed information on this day's activities, refer to the combat debriefing summaries reproduced here.

This Sea Fury (WE790) from HMAS Sydney *was forced to divert to a US base in Korea after being damaged in action. Note the Royal Navy markings.*

SECRET

DE-BRIEFING REPORT

SYDNEY CAG HMAS SYDNEY 11 OCT 51 SERIAL 19
0630 - 0900 EVENT 'A'

1. MISSIONS

(a)	7 x Fury	8 R/P and Cannon	Strike
(b)	3 x Fury	8 R/P and Cannon	Air Spot
(c)	2 x Fury	Cannon	CAP

2. WEATHER

Perfect over target area.

3. REPORTS

(a) 7 x Fury

The aircraft rocketted the reported bivouac area at DU000147. No activity was seen but rockets fell in the target area destroying at least two houses. A large fire was started and smoke was towering to 5,000 ft when the aircraft returned to base.

A reconnaissance was then made for a reported 155 mm gun at DU070079. Although emplacements were sighted they all appeared to be very empty.

A search was then made for a T34 tank which had been located on aerial photos taken 7 Oct at DU048088. The tank had moved and was not found.

An A/R was made and what appeared to be a newly prepared defensive position was found at DU010056. This was straffed. Another newly prepared defensive position at CU986143 was also straffed.

No FLAK was observed.

(b) 3 x Fury

(i) One Fury air spotted for USS "New Jersey" with the object of destroying a concentration of troops at CU996064. Communications were good but procedure only fair. Shot from her 16" guns fell in the target area and destroyed at least ten houses.

(ii) One Fury air spotted for HMS "Belfast" directing her guns at a Bivouac area at CU990159. Communications and procedure were good. Ranging was not accurate and the shoot was abandoned.

(iii) One Fury allocated for an air spot with USS "Hansom" was unable to establish communications until it was too late for a shoot to be carried out.

4. INTELLIGENCE

Two newly prepared defensive positions were sighted at DU010056 and CU986143. They did not appear to be occupied.

5. DAMAGE

DE-BRIEFING REPORT

SYDNEY CAG	HMAS SYDNEY	11 Oct 51	SERIAL 20
		0745 - 0930	EVENT 'B'

1. MISSIONS

(a)	6 x Fireflies	2 x 500lb bombs and Cannon	Strike
(b)	2 x Firefly	2 x 500 lb bombs and Cannon	Air Spot
(c)	2 x Fury	Cannon	CAP

2. WEATHER

Perfect over target area.

3. REPORTS

(a) 6 x Fireflies

After a thorough search the aircraft were unable to locate their primary bombing target of two tungsten mines in the area of DT215933 (P62 and 63). The mission then attacked the dump target of a troop bivouac area at CU990159 (P51). All bombs fell in the village destroying at least six houses. No FLAK was encountered.

(b) 2 x Firefly

(i) One firefly briefed to shoot with USS New Jersey adjusted her fire at a supply dump at DT057999. Communications and procedure were good. Three of her 16" guns fired salvos accurately into the target area, where some ten buildings were left in ruins.

(ii) One Firefly spotted for HMS Belfast whose procedure and communications were again good. The target was a troop concentration at CU971129 (P70). Ranging with phosphorous was accurate but fire for effect with HE was off target. Shortage of time prohibited further corrections and the shoot was abandoned.

4. INTELLIGENCE

The Tungsten mines given as a target are either very small or do not exist at the pin point map reference given.

5. DAMAGE

At least sixteen houses were destroyed.

(Sgd) M.B. SIMKIN Major
DE-BRIEFING OFFICER

SECRET

DE-BRIEFING REPORT

SYDNEY CAG	HMAS SYDNEY	11 OCT 51	SERIAL 21
		0915 - 1115	EVENT 'C'

1. MISSIONS

(a)	7 x Fury	8 R/P and Cannon	Strike
(b)	3 x Fury	.8 R/P and Cannon	Air Spot
(c)	2 x Fury	Cannon	CAP

2. WEATHER - Perfect.

3. REPORTS

(a) 7 x Fury

Village of CHONCHON at DU0309 (P33) now nearly destroyed. This strike knocked over about seven houses. A Pillbox at the south end of a village CHILBORI DU9523 was rocketted and the village straffed. The Pillbox was hit but appeared only very slightly damaged. Several straffing runs were made on the village and the boats along the foreshore, some boats were possibly damaged. A village at CU983136 was straffed. Also a village at CU9129, a few huts being hit.

Troops dug in at hill 78 DT918209 were straffed, no definite damaged was observed as the positions were under trees.

(b) 3 x Furies

HMS Concord's guns were directed onto the town of SONGJON-NI DU002147 and threw sixteen 4.5" shells into the target area, spreading a fire which was already burning. Concord was then switched to a village at CU983136. All the fire for effect fell in the village destroying at least six houses. One large building was very badly damaged.

(ii) The second Fury had the privilege of shooting with USS New Jersey, the target was the village of KORIM-NI (P57) which was nearly destroyed, 6 huts were knocked down and a dozen damaged.

(iii) The last air spotting plane shot with Belfast the target being a gun position at CU006163 (P65) the rounds fell in the general target area but the pilot reported that no definite damage had been done. The last shoot was with USS Hanson shooting onto two positions on the West edge of a village at DU134098 the fire was very spread out but includ[illegible] target area, 4 houses were destroyed and four damaged.

4. INTELLIGENCE

Approximately six men were seen at HUNGUN-NI DU026135.

A power house was located at CU9307.
Troops dug in were observed at Hill 78 DT918209.
Another power house observed at DT018983.

5. DAMAGE

1 ox cart destroyed.
17 houses destroyed, 4 damaged. 1 large building damaged.
6 huts destroyed, 12 huts damaged.

SECRET

6. FLAK

Light flak observed from DU046023 and from trig point 87-DU096082. SAA fire also came from CU988146. One Fury was hit in the wing by small arms fire, the pilot did not realise he was hit until he landed on. Apparent rifle fire was fired from the side of a large house at DU034102.

(Sgd)-G.H. HARDCASTLE Major
DE-BRIEFING OFFICER

The Australian Fireflies had great success against North Korean railway bridges, as this photograph shows. The Koreans rebuilt them very quickly, only to have them knocked down again. This process continued! (via Norman Lee)

SECRET

DE-BRIEFING REPORT

SYDNEY CAG — HMAS SYDNEY — 11 OCT 51 — SERIAL 22

1000-1245 — EVENT 'D'

1. MISSIONS

(a)	6 x Firefly	2 x 500lb Bombs and Cannon	Strike
(b)	2 x Firefly	2 x 500lb Bombs and Cannon	Air Spot
(c)	2 x Fury	Cannon	CAP

2. WEATHER

Perfect.

3. REPORTS

(a) 6 x Firefly

All Fireflies dive bombed the town of TONGCHOW DU050060 most of the bombs falling in the area of the cross roads at DU051050. Fires were burning with dark orange flame an hour after the attack. A building on a Railway Station at DT2594 was straffed.

(b) 2 x Firefly

(i) The first shoot was with USS New Jersey shooting onto a reported tropp concentration at DU023134 (P73). Approximately 15 rounds were fired with the majority hitting the general target area. Communications were intermittent.

(ii) The second Firefly spotted for Belfast shooting at the village of CHANGDONG-NI CU973156. The first ranging round fell in the dead centre of the village starting a number of fires. The second shoot was with Concord, the target being the village CHIGUNG-NI CU027146. This shoot was not very successful, quite a number of the rounds falling into the water. Five rounds fell into the village.

4. INTELLIGENCE

The town of TONGCHON had been badly damaged before the attack took place.

5. DAMAGE

Five houses set on fire. Several houses destroyed.

6. FLAK

No flak reported from the town of TONGCHON. About two bursts of approximately 40mm. flak were fired, bursting between points A and B. This possibly could have been the contentious air burst shell from Belfast when clearing guns from previous flak suppression.

(Sgd) G.H. HARDCASTLE Major
DE-BRIEFING OFFICER

SECRET

DE-BRIEFING REPORT

SYDNEY CAG | HMAS SYDNEY | 11 OCT 51 | SERIAL 23
1230-1430 | EVENT "E"

1. MISSIONS

(a) 14 x Fury	8 x R/P and Cannon	Strike
(b) 3 x Fury	8 x R/P and Cannon	Air Spot
(c) 2 x Fury	Cannon	CAP

2. WEATHER

Perfect.

3. REPORTS

(a) 14 x Fury

(i) Eight Furies rocketted a reported supply dump in a village at DT 108965. No activity was seen but the rockets fell in the target area, several fires were started from what appeard to be supply dumps in the trees next to the village.

The reported Command Post at DT 098966 was straffed. Only houses were sighted at the map reference - these were well covered by the strike and a large fire was left burning.

(ii) Six Furies rocketted and straffed a village at DU 125088 (P34) reported to contain 150 troops. Excellent coverage was obtained, many fires were started, including one large secondary one.

(b) 2 x Fury

(i) One Fury carried out Air Spot with USS New Jersey. Communications and procedure were good. The first target was a village at CU 952112 (P71)reported to contain a troop concentration. The village does now no longer exist! The village containing troops and supplies at DU 022024 (P56) was the next to suffer. Excellent results were obtained and all rounds fell on the target, destroying seven houses.

(ii) One Fury carried out shoots with HMS Belfast and Concord. With HMS Belfast the target was troops on a hill at DU 013113 (P75). Some 30 to 40 troops were seen in trenches and the shoot gave an excellent coverage of the area, many direct hits being made on the trenches. With HMS Concord the target was troops in a village at CU 970149 (P76). The majority of the rounds fell in the area but no actual damage was observed.

4. INTELLIGENCE

(a) Three Automatic AA weapons are sited in emplacements at DU 035006, DU 037008, and DU 039101.

(b) Occupied defensive positions are sited at DU 013113 (P75).

5. DAMAGE

(a) One supply dump damaged.

(b) 18 houses destroyed.

6. FLAK

Light inaccurate FLAK was encountered whilst flying over DU 013113.

SECRET.

DE-BRIEFING REPORT

SYDNEY CAG	HMAS SYDNEY	11 OCT 51	SERIAL 24
		1415-1600	EVENT 'F'

1. MISSIONS

(a)	9 x Firefly	2x500 lb bombs and Cannon	Strike
(b)	2 x Firefly	2 x 500 lb bombs and Cannon	Air Spot
(c)	2 x Fury	Cannon	CAP

2. WEATHER

Perfect over target area.

3. REPORTS

(a) 9 x Firefly

The aircraft bombed a village reported to be a troop concentration with LAA defensives at DU035100. The majority of bombs fell in the village destroying and damaging many houses. Fires were burning long after the attack. During A/R villages at DT1693 and DT2892 were strafed. Tungsten mines in the DT2193 area again proved elusive and not one was sighted.

(b) 2 x Firefly

(i) One firefly spotted for USS New Jersey directing her 16" guns onto two gun positions and one CP in the area CU965065 (P53). The rounds fell in the target area - one house was destroyed and at least one damaged.

(ii) One Firefly spotted for HMS Belfast directing her fire onto troop positions on the ridge CU9810 to DC0011. Many hits were observed.

4. INTELLIGENCE

(a) Light, meagre, inaccurate flak was observed in the area DU0410.

(b) The flak mentioned above has been observed on several occasions previously when it was erroneously thought to have been HMS Belfast clearing her guns or inaccurate flak suppression.

(c) Occupied defensive positions were seen in the area DU0011.

(d) The majority of villages around KOJO were observed to be on fire.

5. DAMAGE

8 houses destroyed and 11 damaged.

(Sgd) M.B. SIMKIN Major
DE-BRIEFING OFFICER

SECRET

DE-BRIEFING REPORT

SYDNEY CAG　　H.M.A.S 'SYDNEY'　　11 Oct 51　　SERIAL 25
1545 - 1700　　EVENT 'G'

1. MISSIONS

(a) 12 x Fury	8 x RP and Cannon	Strike
(b) 3 x Fury	8 x R/P and Cannon	Air Spot
(c) 2 x Fury	Cannon	CAP

2. WEATHER

Perfect.

3. REPORTS

(a) 12 x Fury

Two flights of Furies were briefed to R/P and strafe the ridge between DU 970098 and DU 013119. All pilots reports seeing hundreds and hundreds of enemy troops an estimated hundreds . small store dumps dotted all over the general area of the Eastern slopes of the ridge within a radius of 1 miles of DU 013119. Both troops and dumps were repeatedly rocketted and straffed. Troops had been digging in and the trenches were only partly dug with resulting chaos and confusion on the ground with enemy running in all directions. A conservative estimate of enemy killed or wounded by the Furies is 200.

R/P and straffing attack were also made on two reported enemy positions at DT 076946 and DU 083966. Some movement was observed. During the straffing attack a camouflaged vehicle was set on fire at DT 068936. A fire was started at DT 087968.

(b) 3 x Fury

(i) One Fury shot with Concord firing at a village at DT195982. Two houses were destroyed, one damaged. The road in the village was cut in three places.

(ii) The second Fury used USS New Jersey's 16 inch guns to shoot at troops at DT 076945 (PA 4). Twelve rounds were landed in the hut area, which was severely damaged. The Spotter then adjusted on to trenches and emplacements nearby and another fifteen rounds were landed right into the trench area. No accurate estimate of casualties could be made. One small fire was started in the hut area.

(iii) The Spotter with Belfast fired at a bivouac area DT 112965 (985). The target area generally was well saturated. No specific damage observed.

4. INTELLIGENCE.

One approx 37 m.m. gun observed on the Northern slopes of "Duck Hunt" ridge. On the northern edge of a village at DT 195982 possible stores dump camouflaged with hay was seen.

It would clearly appear that since the strike operation commenced in the KOJO area yesterday morning the enemy in anticipation of a landing have been making feverish attempts to reinforce the area and prepare new defensive positions.

5. DAMAGE

One truck destroyed at DU 016116. One truck set on fire at DT 068936. Two houses destroyed, one damaged.

Estimated 200 enemy killed or wounded.

6. FLAK

SAA Fire observed at DU 002112.

*Replenishment time for HMAS*Sydney*, a task performed by a Royal Fleet Auxiliary vessel in Korea. (Keith Western)*

12-13 October: HMAS *Sydney* arrives at Sasebo (Japan) to refuel and re-arm before undertaking another patrol on the west coast.
14 October: Typhoon 'Ruth' causes damage to the ship and the loss of one Firefly (WB396). With the warning of the approach of the typhoon, the larger ships at Sasebo were ordered to sea in an attempt to avoid the danger. This was British policy, while the American idea was to remain in the shelter of port.

The warning came too late and the ships met the full force of 'Ruth', most sustaining damage of some kind.

The worst of the typhoon was experienced between 1700 hours and midnight, winds exceeding 68 knots (130km/h) at which point *Sydney's* wind recorder failed. Mountainous seas of 45 feet (14m) were also experienced, resulting in the ship pitching and to a greater degree rolling to extreme degrees, up to a frightening 35 deg-rees in the case of the latter.

Damage comprised one Firefly washed overboard and six other aircraft seriously damaged, two of the ship's boats, a forklift truck (along with other items of equipment) were lost and five quickly extinguished electrical fires started due to water entering ventilator ducts and causing short circuits in the ventilator fans. There was a real danger of a major fire during the typhoon when aircrafts' underwing fuel tanks detached and ruptured, filling the ventilation system with Avgas fumes. Injuries to personnel was restricted to cuts and bruises and the captain and navigator both stated it was the worst sea they had ever experienced.
15-16 October: HMAS *Sydney* returned to Sasebo. The damage from typhoon 'Ruth' was estimated at £250,000 ($500,000).
17 October: HMAS *Sydney* on passage to the west coast of Korea for further operations in the Han River area.
18 October: 54 sorties flown and several targets successfully attacked including an underground locomotive repair ship at Haeju. Troop concentrations were attacked by rockets and strafed.

The Fireflies knocked out a span of a rail bridge using dive bombing techniques (more efficient low level bombing methods would soon be introduced) and a successful Airspot was carried out with HMS *Belfast*. A visit was made to the ship by Lt Col Ehrgot (US Army), the officer in charge of guerilla activities on the west coast. Ehrgot's codename was 'Leopard', and sorties conducted on his behalf were coded Leopards.

This operational tour (which lasted until 26 October) saw the Fireflies concentrating mainly on railway bridges and tunnels, while the Sea Furies went after coastal shipping and troop concentrations.
19 October: 53 sorties flown against a variety of targets including a camouflaged gunboat in the Chinnampo area and the factory area at Kyomipo. Three Sea Furies were damaged by flak during an afternoon attack on buildings, one aircraft being hit in the fuel tank, another in the engine and the third in the fin.
20 October: 33 sorties flown, a smaller number due to poor visibility over the targets. Despite peace talks being held, operations continued, the aircrews being directed to attack targets in the Han River area. There were regarded as being few worthwhile targets in that area.

Had these – and other – attempts at peace talks been successful, the codewords 'Brandy Wine' would have been issued, causing an immediate cessation of activity. Until that point was reached, it was considered that continuing to hit the enemy hard provided the best negotiations base!
21 October: 53 sorties flown. Six Sea Furies detailed to attack motor and sail junks near Taehwa-Do, the six aircraft involved carrying rockets. The result of two strikes on these targets was one coastal vessel, three large motor junks and six smaller junks destroyed and 15 other junks damaged.

Sea Furies also carried out the Sydney Air Group's first close army support missions, with the British Commonwealth Division. Initial difficulties were experienced with communications but the four rocket armed aircraft involved attacked enemy troops accurately.
22 October: HMAS *Sydney* refuelled and replenished; no sorties flown.
23 October: 57 sorties flown, 13 of these being associated with the rescue of a USAF Boeing B-29 crew which was down in the sea north of Chinnampo, one of whom was located by a Sea Fury and picked up by an American rescue amphibian, codenamed 'Dumbo'.

Two close air support missions were flown in support of the British Commonwealth Division, controlled by USAF 'Mosquito' (T-6 Texan) aircraft. In one mission, some 75 enemy soldiers were killed.

The Sydney Air Group's planned activities for 22 January 1952, the fourth last day of Korean operations. As it happened, this day became a replenishment day and no operational flying was undertaken. (RAN)

Chaos on HMAS Sydney's *flightdeck with badly damaged Fireflies in the foreground. Both carry Royal Navy titles. (RAN)*

HMS Unicorn, *which acted as a maintenance ferry carrier in Korea, bringing replacement aircraft and stores to the operational ships. via Joe Barr)*

A Sea Fury is recovered aboard HMAS Sydney *while the ship's search and rescue Sikorsky S-51 hovers above. The helicopter was provided by the US Navy and flown by Americans. (RAN)*

Safely on Sydney's *deck, a Sea Fury folds its wings and taxies to its parking space. (RAN)*

Successful operations against other targets on the ground were also carried out, 817 Squadron's Fireflies damaging a rail bridge and causing the collapse of a rail tunnel thanks to accurate low level bombing

24 October: High winds and difficult deck landing conditions restricted the number of sorties to 32. The Fireflies again demonstrated accurate low level skip bombing techniques by placing another 500lb bomb in a rail tunnel, while the search for other members of the USAF B-29 crew downed the day before continued. The Sea Furies attacked suspected troop concentrations and vessels.

25 October: The sortie rate was slightly reduced to 47 due to the ship needing to top up its Avgas supply. Another successful close support mission was carried out with the British Commonwealth Division, again controlled by a T-6 'Mosquito'. The value of these aircraft was quickly being appreciated by the Australian aircrews.

Four Sea Furies were hit by flak, one of which was forced to ditch off Chinnampo. Lt C M Wheatly successfully evacuated his aircraft and was quickly picked by a search and rescue amphibian and taken to Kimpo. This proved an expensive day for the Sea Furies, another aircraft (flown by Lt Cdr J L Appleby, RN, 808 Squadron's commanding officer) being badly hit in a wing and making a precautionary landing at Kimpo.

26 October: An even more expensive day for *Sydney's* aircraft with the loss of a Sea Fury and a Firefly during the course of 58 sorties. The Sea Fury (Sub Lt N W Napstein) was hit by flak and forced down on a mud bank near a friendly island. The pilot was safely picked up by a boat from HMS *Amethyst*.

The second loss, a Firefly, crash landed in the heart of enemy territory near Sariwon north of Haeju after being hit by flak during an attack on a rail tunnel. The pilot (Sub Lt Neil MacMilland) and his observer (CPO P Hancox) then spent an anxious hour and a half squatting in a ditch before being rescued by the ship's helicopter with cover provided by Sea Furies from *Sydney* and Meteors from the RAAF's No 77 Squadron.

Faced with fading light, the decision to send the helicopter was a difficult one, as the downed flyers were some 75 miles (120km) away and surrounded by the enemy. MacMilland and Hancox kept enemy troops at bay with their Owen sub machine guns, while the arrival of the helicopter saw its observer, CPO Gooding, jump from his aircraft and shoot two enemy soldiers who had crept to within 15 metres of the downed Firefly. The helicopter then made its way to Kimpo.

Despite these dramas, successful operations were carried out, the Fireflies destroying a railway bridge and closing both ends of a tunnel. The Sea Furies accounted for some troops, rolling stock, ox carts, stores dumps and sampans as well as flying close support sorties for the troops on the ground.

This day brought to a close HMS *Sydney's* second patrol, its Air Group thus far having flown 654 sorties (474 offensive and 180 defensive) for the loss of three aircraft. A further 23 aircraft had been damaged by flak, most of it small arms fire.

This patrol had itself witnessed 389 sorties and the expenditure of 96,280 rounds of 20mm ammunition, 1,472 rockets, eight 1,000lb bombs and 174 500lb bombs.

27 October – 2 November: The ship is in Kure (Japan) for re-arming and refuelling.

3 November: HMAS *Sydney* departs Kure for its third patrol, again on the west coast of Korea.

4 November: Planning for the operations due to begin tomorrow is undertaken.

5 November: 53 sorties flown but the Air Group suffers its first casualty when Lt Keith Clarkson DFM, a Sea Fury pilot from 805 Squadron is shot down north of the Han River. His aircraft failed to pull out of a strafing run, the result of ground fire hitting the aircraft during the dive.

Three other aircraft were damaged by small arms fire during the day, but a close support mission was successfully flown for the 1st US Cavalry Division, the rocket armed Sea Furies (again under the direction of 'Mosquito' T-6s) putting their ordnance to within 200 yards of friendly troops.

It was noted with interest that all of the bridges and tunnels destroyed during the previous patrol had been repaired.

6 November: 55 sorties flown but their effectiveness was hampered by bad weather. Despite this, a rail bridge and road bridge were destroyed along with several troop shelters. In addition, a successful strike was made on enemy shipping which had been assisting in the invasion of a friendly island at the northern end of the Yellow Sea. Eight vessels were destroyed and others damaged.

The almost immediate repair of bridges by the enemy prompted the use of dropping delayed action bombs as well as quick fused ones. It was hoped this would discourage the overnight repair teams somewhat!

This was also Melbourne Cup day, the running of which fortunately did not coincide with any briefings. One pilot was in the middle of a mission when he was told over the radio he had won £100 ($200) in the ship's Melbourne Cup sweep!

7 November: With snow starting to fall on the battle front, the Air Group flew 53 sorties including another close support mission, this time with the Royal Australian Regiment. Another rail bridge had a span removed and a delayed action bomb was placed on another span.

8 November: 54 sorties flown and two more rail bridges made unserviceable. Air Spots were attempted with the frigate HMAS *Murchison* but as was usually the case with smaller ships, great difficulties were experienced with accuracy, making them suitable only for targets of large area.

9 November: No sorties flown due to ship's refuelling.

10 November: 55 sorties flown including an Army Support mission with 25 Canadian Brigade plus Air Spots with HMS *Belfast*.

11 November: Armistice Day, and appropriately, cease fire negotiations were apparently close to agreement, although as history shows, the war (which at that stage was regarded as being as good as won) would in fact escalate considerably with the commitment of communist China's army and air force to the battle. This action suddenly put the allied forces on the back foot and ensured the war would continue for some considerable time.

An awkward moment for the pilot of Sea Fury VX729 during operations in Korea aboard HMAS Sydney. *(Keith Western)*

Fifty-three sorties flown and by the end of the day the Air Group was able to claim all railways in its area of operations were unserviceable. In addition, a close support mission was flown with 28 Commonwealth Brigade and enemy troops, shipping, vehicles and supply dumps were successfully attacked. One aircraft suffered damage from small arms fire.

12 November: Extremely bad weather restricted the day's sortie tally to only 22, one of which was the landmark 1,000th sortie for the Air Group.

13 November: The final day of this patrol and responsibilities handed over to *Sydney's* relief carrier, USS *Rendova*. Another rail bridge was taken out, a close support mission with 3 Battalion Royal Australian Regiment flown along with an Air Spot with the USS *New Jersey*, the battleship's 16 inch guns destroying troop shelters in a village near Changyon.

14-17 November: HMAS *Sydney* in Sasebo (Japan) replenishing in preparation for its fourth patrol of the Korean War.

18 November: HMAS *Sydney* sails to begin her fourth Korean patrol, this time off the east coast. Other ships involved in the task force are HMS *Belfast* (flag), the destroyers HMAS *Tobruk*, HMS *Constance*, HMCS *Sioux*, NethMS *Van Galen* and USS *Hyman* and three rocket ships. The operations are designed to destroy the installations, defences and communications of the industrial centre and port of Hungnam.

19 November: Briefings for the planned two day attack on Hungnam.

20 November: 58 sorties flown. The large buildings and warehouses of the port were the primary targets and all Fireflies and most Sea Furies carried bombs. At the end of the day, the Australians would claim the destruction of 11 large buildings and the damaging of 31 others.

In addition, Air Spots were carried in conjunction with *Belfast, Tobruk* and *Van Galen*, although these were only moderately successful as the ships were firing from extreme range. Two aircraft damaged by flak.

21 November: 65 sorties flown, most of them similar to the previous day's strikes against Hungnam with the destruction of 29 large buildings and the damaging of a further 21. A rail tunnel was also blocked.

The fact that the bombarding ships came closer to shore meant the Air Spots were more successful than the previous day, although area rather than specific targets were selected. Later in the day *Sydney* departed Hungnam for the west coast of Korea.

22 November: On passage to the west coast.

23 November: Rough seas and very low cloud precluded any flying activity.

24 November: Continued low cloud and snow storms, despite which 36 sorties were flown against targets of opportunity (two spans of a rail bridge were broken) and a close support mission assisting 3 Battalion RAR. Snow fell on the ship!

25 November: Replenishment day and no flying. The rough seas prevented the re-

(opposite and above) HMAS Sydney's *last days of Korean operations in January 1952 were marked by snow storms which sometimes restricted flying. This series of shots gives some idea of the freezing conditions and resultant difficulties. (RAN)*

plenishment ship (RFA *Wave Chief*) from completely refuelling *Sydney*.

26 November: Once again very rough seas inhibited flying, only seven sorties being completed. Yet again it was noted that every rail bridge which had been 'destroyed' or damaged during the last patrol had been repaired.

27 November: Better weather and 60 sorties flown, the Fireflies destroying four rail bridges and one village which was reportedly housing Chinese troops.

The Sea Furies flew a Close Army Support mission with the British Commonwealth Division as well as successfully attacking rolling stock, junks and troop concentrations. Air Spots were carried out with HMShips *Ceylon* and *Mounts Bay*. One aircraft was damaged by flak.

28 November: The last day of HMAS *Sydney's* fourth patrol and 54 sorties were flown. Two more rail bridges destroyed, once again making all rail lines in the ship's area of operation unserviceable. A large troop concentration near Pyongyang was attacked twice, with at least 100 casualties caused.

A close support mission was aborted when the aircraft were on their way to the target due to peace talks in process. Three aircraft damaged by flak.

29-30 November: On passage to Kure for replenishment. By now, *Sydney's* aircraft had flown a total of 1,326 sorties (966 offensive and 360 defensive) for the loss of four aircraft. Forty-three aircraft have been hit by flak.

Due to the relative brevity of this most recent patrol (and the loss of some days due to the weather), only 270 sorties were flown.

1-4 December: HMAS *Sydney* at Kure.

5-6 December: On passage to the west coast of Korea for *Sydney's* fifth patrol of the war. There will be a change of priorities this time, however, to counter the communists who have recently been very active in the islands to the north of the Yellow Sea and have captured all the major ones. It was suspected that the islands of Chodo and Sokto-Ri would be the next to be attacked, so this patrol was aimed at defending those.

7 December: 54 sorties flown and four Sea Furies hit by flak, one of them making an emergency wheels up landing on the beach of the friendly island of Paengyong-Do.

This day saw the Air Group's second casualty. Sub Lt Richard Sinclair's Sea Fury was hit by flak near Chinnampo and although the pilot was able to bail out, he hit the tailplane of his aircraft while doing so and was fatally injured. Sinclair's body was recovered from the sea by a helicopter.

Despite the losses, the day was generally successful with photographic and visual reconnaissance of shipping which may threaten the friendly islands being carried out. The result was the discovery of 109 such vessels. Some 18 small vessels were destroyed or damaged by Sea Fury rocket attacks.

The Fireflies destroyed six rail bridges with the result that four of the seven rail lines in *Sydney's* sphere of influence were unserviceable. Yet again, it became obvious that repairs were being carried out overnight.

8 December: The effects of the temporary 'unserviceability' of several pilots began to manifest itself, although 51 sorties were flown nevertheless. More junks and sampans were destroyed and an unusually large concentration of enemy troops (more than 400) was attacked, of which some 30 were killed.

The Fireflies continued attacks against rail bridges, three more being destroyed and making all but one of the rail lines unserviceable. Three aircraft were hit by flak, a Firefly force-landing on the friendly island Taechong-Do.

9 December: 54 sorties flown, mainly against small vessels and troop concentrations. The result was 11 junks and 19 sampans destroyed or damaged and at least 80 troops killed.

10 December: Once again, 54 sorties were flown, the Fireflies responsible for the dropping of two spans from a bridge on the last serviceable railway line in *Sydney's* area and the destruction of 22 buildings which it was believed housed troops.

The Sea Furies continued attacking junks and sampans with success as well as attacking troop concentrations in the Changyon area with an estimated 134 killed. Three aircraft were hit by small arms fire.

The method used to assess enemy troop casualties was as follows: unless the bodies were actually counted, a rocket must have been seen to strike a known number of troops before they were accepted as being killed. The result was a usually conservative estimate of casualties, particularly when a large concentration of troops was attacked.

11 December: 56 sorties flown, mainly against troops and small vessels, resulting

in at least 110 troops killed and 35 vessels destroyed or damaged. Two aircraft hit by light flak.
12 December: Replenishment day from RFA *Wave Premier*, so no flying.
13 December: 53 sorties flown but at the cost of two Sea Furies shot down and four damaged by flak. Both Sea Fury pilots were rescued by the Paengyong-Do helicopter. One of the pilots came down in enemy territory, these rescues once again demonstrating the willingness of the American helicopter pilots to risk themselves to save the downed airmen.

The day's activities resulted in the destruction of one more rail bridge (bringing the total to 28), 14 troops, several junks and sampans and the inevitable ox carts.
14-15 December: No flying due to heavy swells, high winds and snow storms.
16 December: HMAS *Sydney's* third birthday (since commissioning), during which time she had steamed 114,691.1 nautical miles (212,443km), or the equivalent of 4.6 times around the world.

This day saw the ship's aircraft provide CAP and ASP services for a nine ship convoy bringing troops to Inchon. Consequently, no offensive sorties were flown.
17 December: The final day of *Sydney's* fifth patrol and 55 sorties were flown, mainly against suspected troop concentrations and gun positions in the Chodo-Sokto-Ri area. Five aircraft damaged by flak.
18 December: On passage to Kure for replenishment and re-arming and hopefully a pleasant Christmas!

Sydney's fifth patrol logged 338 sorties and proved to be the most expensive thus far with five aircraft and one pilot lost. A further 20 aircraft were damaged by flak, giving a relatively high 'hit' rate of one for every 12 offensive sorties flown.
19-26 December: HMAS *Sydney* at Kure.
27 December: On the day an armistice was supposed to be signed, the ship sails to begin her sixth patrol, once again on Korea's west coast, in this case relieving the USS *Bandoeing Strait*. On board are 24 Australian and New Zealand officers and men, there to gain experience of carrier operations.
28 December: On passage to the west coast. Tasks on this trip are expected to be similar to the previous patrol: Sea Furies providing daily cover (TARCAP) for shipping in the Chodo-Sokto-Ri area, reconnaissance for attacks on small enemy vessels, interruption of rail communications and strikes on 'Leopard' targets, those provided by friendly guerilla forces on the ground.
29 December: Most of the Sydney Air Group's operational area is by now covered in snow and nearly all of the rivers frozen. North of Chinnampo, the sea was frozen inshore, while pack ice had formed further out. The result was a decrease in enemy shipping movements and more difficult map reading for the aircrews.

Fifty-four sorties were flown on the first day of *Sydney's* sixth patrol, resulting in the destruction of three railway bridges. A Firefly attacking one of the bridges at low level was hit by flak and had to divert to Seoul.
30 December: The Air Group provided a continuous CAP and ASP over two large convoys moving the US 1 Cavalry Division from Inchon to Japan. The sorties tally for the day reached 32, although it is worth noting that the aircrews found these missions to be somewhat boring by comparison with the offensive flying of previous patrols.
31 December: 54 sorties flown with another three rail bridges destroyed by the Fireflies, resulting once again in the cutting of all rail lines in the area. Despite this, the communists continued to rapidly repair the damage, now taken advantage of the frozen and low level of water by building low embankment diversions around the bridges. As these embankments were poor targets and easily repaired, cutting rail traffic completely was becoming increasingly difficult. One Sea Fury damaged by flak.
1 January 1952: The new year was marked with the breakdown of *Sydney's* catapult, this, in conjunction with an emergency landing by a flak damaged aircraft, suspending flying for two hours.

Fifty-one sorties were flown, the main effort going towards supporting friendly guerillas forced to evacuate the island of Yongwi-Do to the south of Ongjin. Furies and Fireflies together attacked troop concentrations, allowing the successful evacuation of the friendly forces.

Photographic sorties had always been carried out by *Sydney's* aircraft, but the snow covered landscape prompted an increase in activity to cover as much of the target area as possible.
2 January: Poor visibility restricted offensive flying. In all, 47 sorties were flown comprising 19 offensive, 14 CAP and ASP, 8 TARCAP and 6 SAR.

The six search and rescue sorties resulted from the loss of *Sydney's* third pilot of the war, Sub Lt Ronald Coleman. Coleman's Sea Fury went missing during a CAP mission, extensive air and sea searches finding no trace of the aircraft or its pilot. Two aircraft were hit by flak during the day.
3 January: 56 sorties flown, the Sydney Air Group passing the 2,000 operational sorties milestone since deployment to Korea during the day.

Most of the offensive sorties were in the Ongjin area where enemy troops were threatening nearby friendly islands. In addition, the Fireflies bagged their 38th rail bridge, the low level bombing techniques proving extremely accurate.

It may be remembered that early in the Korean sojourn, the Fireflies used dive bombing techniques, something the aircraft was not entirely suitable for. In these cases, the Fireflies were dived from 8000 to 3000 feet at an angle of 55 degrees. The use of low level techniques (dropping normally 27 second delayed action bombs so as to allow all aircraft to clear) increased accuracy to a high degree.
4 January: Replenishment from RFA *Brown Ranger* – no flying.
5 January: 54 sorties flown covering the area from Hanchon to the Han River. The emphasis of the Air Group's battle was by now gradually shifting southwards to the Ongjin, Haeju and Yonan areas due to large enemy troop builds-up there and the fact that the sea to the north was rapidly freezing. Anti shipping runs to Hanchon were cancelled. The Fireflies conducted strikes against buildings around Yonan.

One Sea Fury pilot – Lt P Goldrick – had a close shave when his aircraft was hit by flak near Chinnampo. Despite having collected a bullet in the arm, Goldrick brought his aircraft safely back to the carrier. Another Sea Fury also suffered flak damage.

This day also saw Sea Furies carrying out Air Spots with HMShips *Alacrity* and *Mounts Bay* and the Dutch *Van Galen*. As had previously been the case with small warships, this trio's accuracy was poor, with only 30 per cent of shot falling within the target area.
6 January: Rough seas, snow storms and poor visibility resulted in only one mission (comprising 14 sorties) being flown against a suspected troop concentration in Yonan.

With the flight deck covered with slush from snow, HMAS Sydney's *aircraft remain tied down late in the ship's Korean adventure.*

'Sydney' – in snow! (RAN)

At least 24 buildings were destroyed and two aircraft hit by light flak, one of them five times.

The comparative effectiveness of 60lb rocket projectiles and 500lb instantaneously fused bombs against these targets was noted. The rockets caused high casualty rates among those inside the buildings but rarely destroyed them, while the bombs' blast effect flattened the buildings and usually started fires.

7 January: Snow storms prevented flying, frustrating because the weather at the targets was good. HMAS *Sydney* attempted to steam out of the bad weather without success.

8 January: The end of *Sydney's* sixth patrol and on passage to Kure for replenishment and rearmament.

9-14 January: In Kure.

15 January: HMAS *Sydney* leaves Kure for her seventh and last patrol of the Korean war.

16 January: On passage to Korea's west coast.

17 January: Intermittent snow storms allowed only 43 sorties to be flown, with little resistance from ground fire. The day's flying saw the destruction of two more rail bridges, several attacks on ground targets in the Ongjin and Han River areas and the provision of TARCAP for anti invasion shipping near Chodo-Sokto-Ri.

18 January: Good weather and 54 sorties flown including attacks on several targets from Chinnampo to the Han River, the Chodo-Sokto-Ri TARCAP was maintained and a combined Sea Fury and Firefly strike against villages reported to be housing enemy troops resulted in the destruction of at least 19 buildings. One Firefly and two Sea Furies were damaged by flak, the Firefly making a precautionary landing at Seoul.

19 January: 52 sorties flown against rail bridges and other targets in the Ongjin, Ullyul and Han River areas. One Sea Fury hit by flak near Ongjin.

20 January: Despite fog obscuring most of the target area during the day, 55 sorties were flown comprising 36 Sea Fury and 19 Firefly, of which 30 were offensive, 10 were TARCAPS and 15 were defensive.

817 Squadron's Fireflies had what has been described as a 'red letter day', destroying four and damaging one rail bridge, yet again putting all the rail lines in the Air Group's area of operations out of action.

Sea Furies and Fireflies again combined, this time to attack enemy troop concentrations on the recently captured island of Chanonin-Do. Most buildings on the island were destroyed and two Sea Furies were damaged by flak.

21 January: HMAS *Sydney* sailed south to provide a Convoy CAP for US ships taking troops to Inchong. Flying was restricted by thick fog and only 15 sorties were recorded.

22 January: Replenishment day, the ship taking on fuel oil and Avgas from the RFA *Wave Premier*. No flying.

23 January: 53 sorties flown resulting in the destruction of three more rail bridges by the Fireflies, bringing their tally for the war to 47. Attacks were also made on villages in which anti aircraft guns were housed and the Sea Furies discovered a new type of target when they attacked and damaged two mobile cranes in the Kyomipo factory yards. One Sea Fury was damaged by light flak near Ongjin.

24 January: Flying was cancelled in the middle of the day due to poor weather, but before that 21 sorties were flown, both the Sea Furies and Fireflies attacking troop concentrations in and around Yonan and Segori with the destruction of numerous buildings recorded. As it turned out, these were the last operational sorties flown by HMAS *Sydney's* aircraft in Korea.

25 January: The Sydney Air Group's final operational day was something of an anti climax as no flying was possible due to high winds, rough seas and freezing temperatures (-8 deg C) which resulted in the aircraft on deck and the deck itself being covered with a sheet of ice.

26 January: On passage to Sasebo.

27 January: HMAS *Sydney* departed Sasebo bound for Australia via Hong Kong: "Without regret", according to the Air Group's diary.

28-29 January: Preparations for handing over to HMS *Glory* which would resume operations in Korea. A substantial amount of equipment had to be transferred to the British aircraft carrier in Hong Kong including the following items: 50,000 maps and indexes, 5,500 aerial photographs and associated plots, logs and indices, 57 mosaics, 17 'made up' rail bridge briefing sheets, 1 interdiction map, 20 special debriefing maps, 90 bombardment spotting maps, 1 set of plastic relief maps, 120 town plans, 620 silk maps, 85 Far East target dossiers, 50 sets of fluorescent panels (for rescue), 50 'walkie talkies', 67 blood chits, 50 survival booklets, 10 message dropping bags, 100 bombardment spotting cards, 50 sets of arctic underwear, 12 Owen guns and ammunition, 1 flak trace, 1 friendly island trace, 1 general military situation trace, 1 bombline and neutrality zone trace plus assorted secret files and security publications.

30 January: HMAS *Sydney* and HMS *Glory* met outside Victoria Harbour, Hong Kong. Six Fireflies were flown to *Glory*. The ships were moored in Victoria Harbour later in the day when the transfer of stores and equipment was begun.

31 January: The formal conclusion of HMAS *Sydney's* Korean tour of duty, during which time the ship had spent 64 days on station in the operational area. Of these, 9.5 days were spent replenishing or on passage between the east and west coasts and bad weather caused the loss of 11.7 flying days.

Messages to the ship at the end of her tour included one from the Admiral Commanding the US 7th Fleet: "Departure of *Sydney* is met with mixed feelings of regret and joy. Setting records of getting planes in the air in all conditions of weather, striking troop concentrations and supply lines and furnishing air spots for United Nations ships are only a few of your many achievements. Fighting side by side with other United nations units, *Sydney* has upheld [the] highest traditions of [the] Commonwealth Navy. With great respect for the proven fighting ability of this fine ship and its embarked Air Group, I extend to all hands our heartfelt thanks for a job well done. May you have a safe and happy journey to a well deserved rest."

HMAS *Sydney* departed for home, but later the following year she would be back in Korean waters, although this time without sending her aircraft off in anger.

Uneasy Truce

After completing her Korean tour of duty, HMAS *Sydney* returned to Australia for operations in home waters. In March 1953 she was again away on an international cruise, this time travelling to the United Kingdom via the Suez Canal with the Coronation Contingent on board. The ship returned to Australia in August, travelling via the USA (Baltimore) and Pearl Harbour. While in Britain, 37 more Fireflies and 12 additional Sea Furies were taken on board.

Meanwhile, in Korea, an armistice had been signed, formally ending the Korean war on 27 July 1953 and leaving the opposing sides occupying much the same territory they had before hostilities began. Korea was still divided into two halves, North and South.

At that stage no peace settlement had been signed, leaving an uneasy peace which required armistice patrols to be carried out. The Royal Australian Navy contributed to this in the second half of 1953 and into the following year, committing HMAShips *Tobruk, Arunta, Condamine, Culgoa, Shoalhaven* and *Murchison* to the peace keeping effort.

Also allocated to the task was HMAS *Sydney*, under the command of Captain G C Oldham and with two squadrons of Sea Furies and one of Fireflies aboard. The squadrons differed this time, only 805 Squadron with its Sea Furies (now under the command of Lt Cdr A J Gould) carrying over from the previous tour. The other Sea Fury squadron was No 850 (Lt Cdr P M Austin RN) while the Firefly unit was 816 Squadron, commanded by Lt Cdr W G Herbert.

Sydney arrived in Korean waters in November 1953 and departed in June 1954. This time there was no combat flying and no losses.

HMAS *Vengeance* was also marked down for Korean service in 1954 but in the event visited Japanese waters only, from where she took home the RAAF's No 77 Squadron after a lengthy period of overseas service which had begun during World War II, continued with the British Commonwealth Occupation Forces in Japan and ended with the Korean conflict.

RAN HAWKER SEA FURIES

Abbreviations: RN - Royal Navy; RAN – Royal Australian Navy; FAA – Fleet Air Arm; NAS – Naval Air Station; w/o – written off; deliv – delivery; soc – struck off charge.

Notes: The majority of RAN Sea Furies were sourced from the United Kingdom and delivered via HMAS *Sydney* or via HMAS *Vengeance* (10 aircraft – WM479-482, WM490, WJ279, WJ284, WJ294, WJ297, WJ299).

Eight Sea Furies were transferred from RN stocks in Korea (WE686, WE791, WE795-797, WE799, WF591, WF593), and ten came from the Royal Navy's NAS at Sembawang, Singapore (WG627, WG628, WH581, WH586, WH588, WH589 aboard HMAS *Sydney*, plus WG630, WH583, WH587, WH590 on *Vengeance*).

Two aircraft (VX627 and VX661) were delivered by merchant vessel.

Serial	Model	Deliv	Disposal/Remarks
TF925	F.10	05/49	converted to components 11/49
VW232	FB.11	05/49	to training aid 03/51, extant Naval Aviation Museum Nowra NSW as 'VX730'
VW622	FB.11	05/49	crashed Nowra NSW 07/52
VW623	FB.11	05/49	extant Naval Aviation Museum, Nowra NSW
VW624	FB.11	05/49	converted to components 01/54
VW625	FB.11	05/49	sold to T Carr & Co 03/59 03/59
VW626	FB.11	05/49	sold to Fitzgibbon & Garthon 06/57
VW627	FB.11	05/49	converted to components 02/52
VW629	FB.11	05/49	fire fighting training NAS Nowra, w/o 10/65
VW630	FB.11	05/49	sold 09/60
VW631	FB.11	05/49	sold to A G Simms 10/56
VW632	FB.11	05/49	sold to T Carr & Co 03/59
VW634	FB.11	05/49	fire fighting training NAS Nowra, w/o 06/62
VW635	FB.11	05/49	sold 11/56
VW636	FB.11	05/49	sold 11/56
VW637	FB.11	05/49	sold 09/60
VW638	FB.11	05/49	converted to components 10/55
VW639	FB.11	05/49	converted to components 11/55
VW640	FB.11	05/49	converted to components 08/53
VW642	FB.11	05/49	sold 09/60
VW643	FB.11	05/49	sold to T Carr & Co 03/59
VW644	FB.11	05/49	sold 11/56
VW645	FB.11	05/49	sold to Fitzgibbon & Garthon 06/57
VW646	FB.11	05/49	crashed Nowra NSW 11/50, engine failure
VW647	FB.11	05/49	soc 11/59, extant Camden Aviation Museum NSW
VW648	FB.11	05/49	crashed in sea 07/51
VW660	FB.11	05/49	sold to Fitzgibbon & Garthon 06/57
VX627	FB.11	03/50	sold 10/56
VX661	FB.11	05/50	sold 11/56
VX707	FB.11	12/50	fire fighting training NAS Nowra, w/o 10/65
VX724	FB.11	12/50	converted to components 07/54
VX725	FB.11	12/50	sold to A G Simms 10/56
VX726	FB.11	12/50	crashed in sea 03/55
VX727	FB.11	12/50	fire fighting training NAS Nowra, w/o 03/62
VX728	FB.11	12/50	lost on operations Korea 10/51
VX729	FB.11	12/50	converted to components 12/55
VX730	FB.11	12/50	sold 03/59
VX748	FB.11	12/50	converted to components 12/51
VX749	FB.11	12/50	fire fighting training NAS Nowra, w/o 06/62
VX750	FB.11	12/50	converted to components 01/56
VX751	FB.11	12/50	forced landing Cape Gloucestor 09/52 and w/o
VX752	FB.11	12/50	fire fighting training NAS Nowra, w/o 06/62
VX753	FB.11	12/50	crashed Nowra NSW 09/54
VX754	FB.11	12/50	crashed in sea 05/51
VX755	FB.11	12/50	converted to components 02/56
VX756	FB.11	12/50	fire fighting training NAS Nowra, w/o 10/65
VX757	FB.11	12/50	sold 09/60
VX758	FB.11	12/50	crashed in sea 04/55
VX759	FB.11	12/50	sold 11/56
VX760	FB.11	12/50	sold to A G Simms 10/56
VX761	FB.11	12/50	crashed in sea 09/51
VX762	FB.11	12/50	crashed after mid air collision 05/53
VX763	FB.11	12/50	converted to components 02/56
VX764	FB.11	12/50	lost on operations Korea 01/52
WE673	FB.11	12/50	sold to Fitzgibbon & Garthon 06/57
WE674	FB.11	12/50	lost on operations Korea 11/51
WE675	FB.11	12/50	sold to T Carr & Co 03/59
WE676	FB.11	12/50	converted to components 02/56
WE677	FB.11	12/50	fire fighting training NAS Nowra, w/o 04/62
WE678	FB.11	12/50	sold 09/60
WE679	FB.11	12/50	lost on operations Korea 11/51
WE686	FB.11	12/51	lost on operations Korea 01/52
WE791	FB.11	10/51	sold 09/60
WE795	FB.11	11/51	lost on operations Korea 01/52
WE796	FB.11	10/51	lost on operations Korea 11/51
WE797	FB.11	10/51	lost on operations Korea 01/52
WE799	FB.11	10/51	sold to A G Simms 10/60
WF591	FB.11	10/51	sold to T Carr & Co 03/59
WF593	FB.11	10/51	sold 09/60
WG627	FB.11	03/52	sold 09/60
WG628	FB.11	03/52	sold to T Carr & Co 03/59
WG630	FB.11	03/52	soc 11/59, extant Naval Aviation Museum, Nowra NSW
WH581	FB.11	03/52	sold to T Carr & Co 03/59
WH583	FB.11	03/52	sold to A G Simms 10/56
WH586	FB.11	03/52	crashed Nowra NSW 04/56 and w/o
WH587	FB.11	03/52	sold 09/63, extant with Ellsworth Getchell, San Jose, California USA, airworthy
WH588	FB.11	03/52	sold 09/63, extant with Lloyd A Hamilton, Santa Rosa, California USA, airworthy
WH589	FB.11	03/52	sold 09/63, components extant in composite R-4360 powered 'Furias' racer, Lloyd A Hamilton, Santa Rosa, California USA
WH590	FB.11	03/52	fire fighting training NAS Nowra, w/o 10/56
WJ279	FB.11	01/53	sold to T Carr & Co 03/59
WJ284	FB.11	01/53	sold 09/60
WJ294	FB.11	01/53	converted to components 07/54
WJ297	FB.11	01/53	sold to T Carr & Co 03/59
WJ299	FB.11	01/53	fire fighting training NAS Nowra, w/o 06/62
WM479	FB.11	01/53	fire fighting training NAS Nowra, w/o 06/62
WM480	FB.11	01/53	sold to A G Simms 10/56
WM481	FB.11	01/53	sold to T Carr & Co 03/59
WM482	FB.11	01/53	sold to A G Simms 10/56
WM490	FB.11	01/53	sold to A G Simms 10/56
WZ642	FB.11	01/54	converted to components 12/55
WZ643	FB.11	01/54	sold to T Carr & Co 03/59
WZ644	FB.11	01/54	crashed Nowra NSW 07/55
WZ645	FB.11	01/54	converted to components 03/58
WZ646	FB.11	01/54	crashed Nowra NSW 10/55
WZ647	FB.11	01/54	sold to Fitzgibbon & Garthon 06/57
WZ648	FB.11	01/54	sold to A G Simms 10/56
WZ649	FB.11	01/54	fire fighting training NAS Nowra, w/o 07/58
WZ650	FB.11	01/54	sold to T Carr & Co 03/59
WZ651	FB.11	01/54	sold to T Carr & Co 03/59
WZ652	FB.11	01/54	sold 11/56

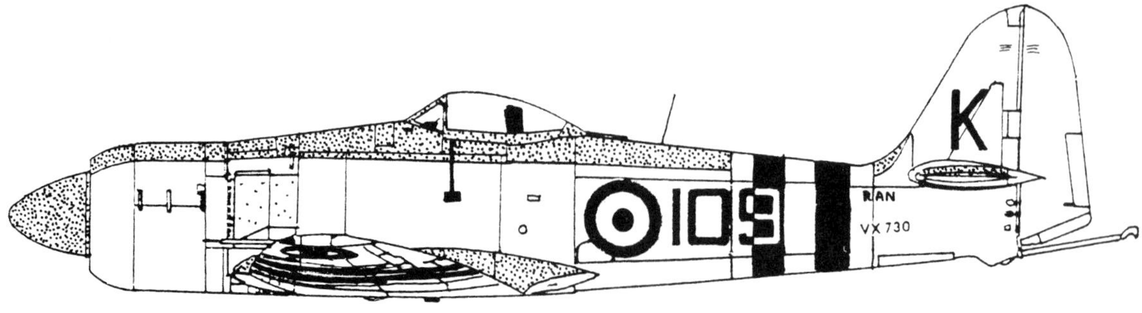

Sea Fury VX730/109K aboard HMAS *Sydney*, Korea, 1951
Standard scheme of extra dark sea grey upper surface, including spinner, with sky fuselage and undersurface. Black numbers, serial and code. Black and white stripes encircle fuselage and wings.

Sea Fury WE679/162K aboard HMAS *Sydney*
Standard scheme. Sky spinner.

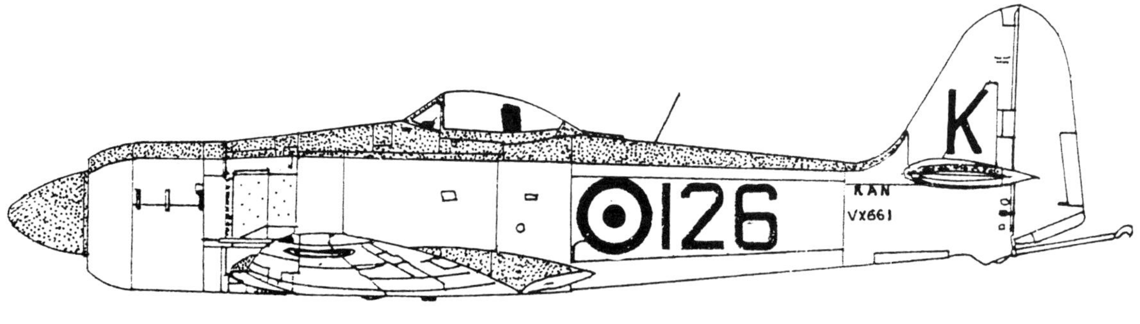

Sea Fury VX661/126K aboard HMAS *Sydney*
Standard scheme.

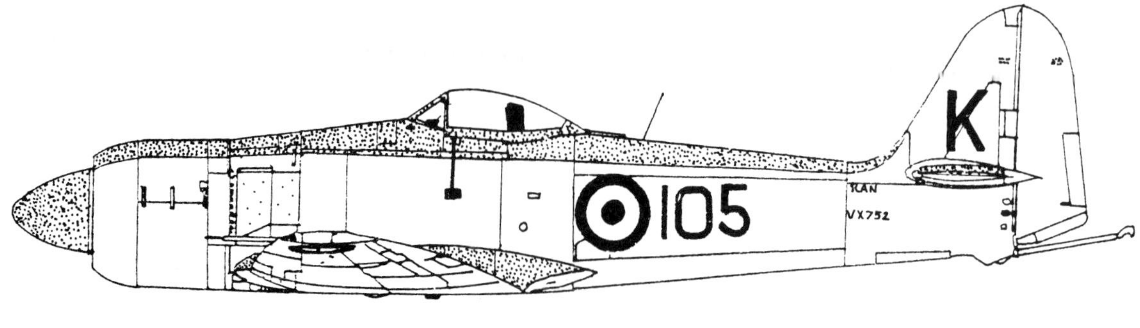

Sea Fury VX752/105K aboard HMAS *Sydney*
Standard scheme.

Sea Fury WH587/105
Standard scheme. Sky spinner.

Sea Fury WH590 used for aerobatic displays
Overall silver. Red canopy frame, spinner, leading edge of fin and wing tips. Black lettering and serial. Black NAVY under right wing. Black 04 on main undercarriage doors.

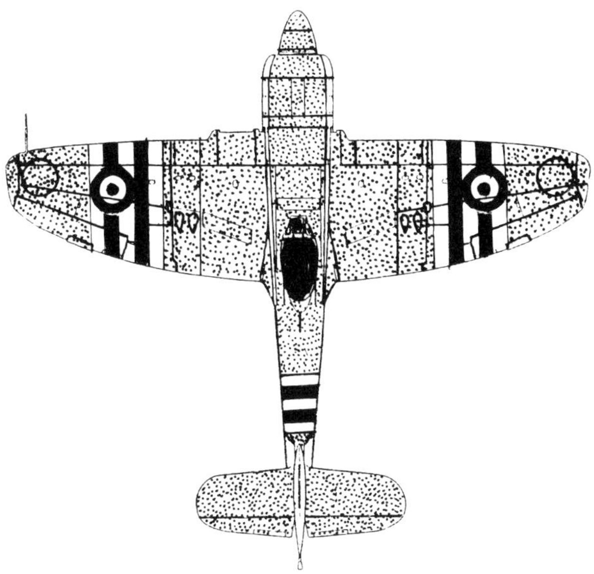

Upper surface, showing location of roundels and the black and white stripes applied during Korean War operations to VX730. The location and size of the roundels remained unchanged throughout the Sea Fury's operational service.

RAN FAIREY FIREFLIES

Abbreviations: w/o – written off; cvtd – converted; RN – Royal Navy; RAN – Royal Australian Navy; FAA – Fleet Air Arm; NAS – Naval Air Station.

Notes: Although most aircraft were obtained direct from the UK and shipped to Australia aboard either HMAS *Sydney* (mainly) or HMAS *Vengeance*, several aircraft were transferred from the Royal Navy in Korea. These were Firefly AS.5s WB316, WB358, WB393 and WB396. Others were obtained from the Royal Navy's NAS at Sembawang, Singapore, these being AS.5s VT440, WB271, WB303 and WB306.

HMAS *Vengeance* was responsible for the delivery of WD826, WD842, WD846, WD855, WD866, WD869, WD882, WD884, WD887, WD891-93, WD896, WD898, WD901, WD906, WJ109, WJ112-13 and WJ121. WB292 and WB338 were delivered by merchant vessel.

Serial	Mark	Deliv	Disposal/Remarks
VG989	FR.4	05/49	converted to components 11/49
VH122	FR.4	05/49	fire fighting training NAS Nowra, w/o 08/58
VT440	AS.5	03/52	cvtd trainer, sold 09/56
VT500	AS.5	05/49	converted to components 12/55
VT501	AS.5	05/49	sold 03/59
VT502	AS.5	05/49	cvtd trainer, fire fighting training NAS Nowra, w/o 06/62
VT503	AS.5	05/49	fire fighting training NAS Nowra, w/o 07/58
VT504	AS.5	05/49	sold to A G Simms 01/58
VX371	AS.5	05/49	crashed at sea 02/51
VX372	AS.5	05/49	sold to T Carr & Co 03/59
VX373	AS.5	05/49	cvtd trainer, sold T Carr & Co 03/59
VX374	AS.5	05/49	sold to T Carr & Co 03/59
VX375	AS.5	05/49	sold to T Carr & Co 03/59
VX376	AS.5	05/49	sold to T Carr & Co 03/59
VX377	AS.5	05/49	sold to T Carr & Co 03/59
VX378	AS.5	05/49	sold to A G Simms 10/56
VX379	AS.5	05/49	converted to components 09/50
VX380	AS.5	05/49	sold to T Carr & Co 03/59
VX381	AS.5	05/49	crashed into sea after collision with Firefly WD887 over Jervis Bay NSW 11/56
VX382	AS.5	05/49	converted to components 09/50
VX383	AS.5	05/49	converted to components 11/55
VX384	AS.5	05/49	sold to Fitzgibbon & Garthon 06/57
VX385	AS.5	05/49	cvtd trainer, converted to components 07/54
VX386	AS.5	05/49	sold to A G Simms 04/59
VX387	AS.5	05/49	sold to Fitzgibbon & Garthon 06/57
VX388	AS.5	05/49	cvtd target towing, sold Historic Aircraft Preservation Society 07/66, extant Camden Aviation Museum NSW
VX389	AS.5	05/49	converted to components 02/56
VX390	AS.5	05/49	sold to A G Simms 01/58
WB271	AS.5	03/52	cvtd target towing, purchased for RN FAA Museum Yeovilton UK 07/66, extant with FAA Historic Flight, Yeovilton
WB292	AS.5	03/50	sold to T Carr & Co 03/59
WB303	AS.5	05/54	converted to components 02/56
WB306	AS.5	03/52	fire fighting training NAS Nowra 05/68
WB316	AS.5	03/52	lost on operations Korea 01/52
WB338	AS.5	05/50	converted to components 04/53
WB358	AS.5	10/51	lost on operations Korea 01/52
WB371	AS.5	12/50	converted to components 11/55
WB393	AS.5	09/51	lost on operations Korea 11/51
WB396	AS.5	10/51	lost from HMAS *Sydney* during typhoon, Korea 10/51
WB423	AS.5	12/50	converted to components 10/57
WB505	AS.6	12/50	sold to Fitzgibbon & Garthon 06/57
WB506	AS.6	12/50	sold to T Carr & Co 03/59
WB507	AS.6	12/50	sold to T Carr & Co 03/59
WB508	AS.6	12/50	sold to T Carr & Co 03/59
WB509	AS.6	12/50	sold 01/58
WB510	AS.6	12/50	converted to components 02/56
WB516	AS.6	12/50	converted to components 07/54
WB517	AS.6	12/50	sold to T Carr & Co 03/59
WB518	AS.6	12/50	cvtd target towing, sold Mr C Beltrame 07/66, extant Classic Aviation, Bankstown NSW under restoration to fly
WB519	AS.6	12/50	crashed at sea 06/52
WB520	AS.6	12/50	sold to Fitzgibbon & Garthon 06/57
WB521	AS.6	12/50	converted to components 07/54
WB522	AS.6	12/50	crashed at sea 02/52
WB523	AS.6	12/50	crashed at sea 02/52
WD824	AS.6	12/50	sold to T Carr & Co 03/59
WD825	AS.6	12/50	sold 09/60
WD826	AS.6	03/53	cvtd target towing, to instructional airframe 1964, extant RAN Historic Flight, Nowra NSW, airworthy.
WD827	AS.6	12/50	sold Australian Air League 11/56, extant Moorabbin Air Museum Vic
WD828	AS.6	12/50	cvtd target towing, sold Aust Aircraft Restoration Group 07/66, flown 1984-87 before forced landing, extant Griffith NSW
WD829	AS.6	12/50	converted to components 10/57
WD830	AS.6	12/50	sold to Fitzgibbon & Garthon 06/57
WD831	AS.6	12/50	converted to components 02/56
WD832	AS.6	12/50	converted to components 02/56
WD833	AS.6	12/50	sold 09/60, extant with Henry J Schroeder, Danville Illinois under rebuild to fly
WD834	AS.6	12/50	sold to T Carr & Co 04/59
WD835	AS.6	12/50	sold to T Carr & Co 04/59
WD836	AS.6	12/50	sold to Fitzgibbon & Garthon 10/56
WD837	AS.6	12/50	fire fighting training NAS Nowra, w/o 06/62
WD838	AS.6	12/50	sold to A G Simms 01/58
WD839	AS.6	12/50	sold to Fitzgibbon & Garthon 06/57
WD840	AS.6	12/50	cvtd target towing, sold Aircraft Engine Overhaul Service 07/66, extant Lone Star Museum, Galveston Texas USA pending restoration
WD841	AS.6	08/53	fire fighting training NAS Nowra, w/o 06/62
WD842	AS.6	03/53	sold to Fitzgibbon & Garthon 06/57
WD843	AS.6	08/53	sold to T Carr & Co 04/59
WD846	AS.6	03/53	sold to Fitzgibbon & Garthon 06/57
WD852	AS.6	08/53	sold to T Carr & Co 04/59
WD855	AS.6	03/53	sold to T Carr & Co 03/59
WD856	AS.6	08/53	sold to T Carr & Co 04/59
WD861	AS.6	08/53	sold 09/61
WD866	AS.6	03/53	sold to Fitzgibbon & Garthon 06/57
WD867	AS.6	08/53	sold to T Carr & Co 04/59
WD869	AS.6	03/53	crashed near Kiama NSW 03/57
WD870	AS.6	08/53	sold to T Carr & Co 04/59
WD879	AS.6	08/53	sold to Fitzgibbon & Garthon 06/57
WD882	AS.6	03/53	sold to Fitzgibbon & Garthon 06/57
WD883	AS.6	08/53	sold to T Carr & Co 04/59
WD884	AS.6	03/53	sold to T Carr & Co 03/59
WD885	AS.6	08/53	sold to T Carr & Co 04/59
WD887	AS.6	03/53	crashed after collision with Firefly VX381 over Jervis Bay NSW 11/56
WD891	AS.6	03/53	crashed at sea 09/55
WD892	AS.6	03/53	sold to Fitzgibbon & Garthon 06/57
WD893	AS.6	03/53	sold to T Carr & Co 03/59
WD895	AS.6	08/53	sold to T Carr & Co 04/59
WD896	AS.6	03/53	fire fighting training NAS Nowra, w/o 06/62
WD897	AS.6	08/53	sold to A G Simms 10/56
WD898	AS.6	03/53	sold to Fitzgibbon & Garthon 06/57
WD899	AS.6	08/53	sold to T Carr & Co 04/59
WD901	AS.6	03/53	cvtd target towing, sold Casula Auto Wreckers NSW 07/66, flown by Canadian Warplane Heritage 1971-77, crashed into Lake Ontario 09/77
WD903	AS.6	08/53	sold to T Carr & Co 04/59
WD906	AS.6	03/53	sold to A G Simms 01/58
WD914	AS.6	08/53	sold to T Carr & Co 03/59
WD915	AS.6	08/53	sold to T Carr & Co 03/59
WD921	AS.6	08/53	sold to T Carr & Co 04/59
WH627	AS.6	08/53	fire fighting training NAS Nowra 08/58
WH632	AS.6	08/53	sold Aust Air League 06/60, Canadian Warplane Heritage 06/78, reflown 04/92, crashed on t-o Victoria BC05/92
WJ109	AS.6	03/53	cvtd target towing, extant NAS Nowra NSW
WJ112	AS.6	03/53	sold to T Carr & Co 03/59
WJ113	AS.6	03/53	sold to T Carr & Co 03/59
WJ121	AS.6	03/53	sold to Fitzgibbon & Garthon 06/57

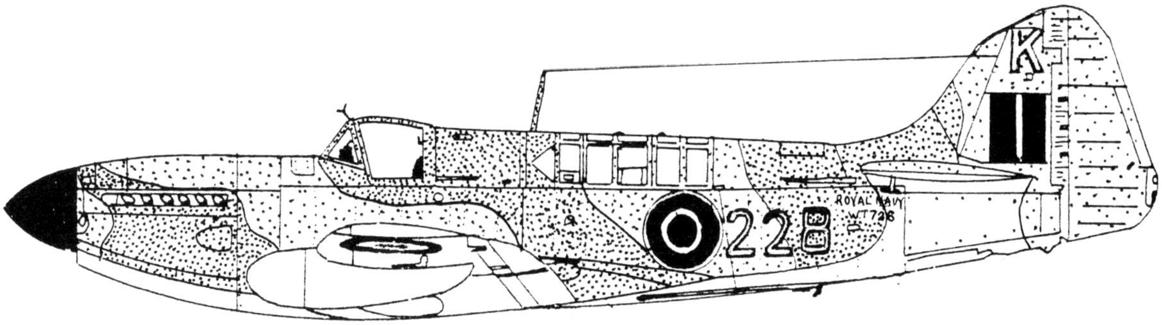

Firefly WT726/228K of 816 Sqn aboard HMAS *Sydney*, 1948
Extra dark sea grey and dark slate grey upper surface, with sky undersurface. White 228 and K. Black serial and spinner. Yellow surround on fuselage roundels only.

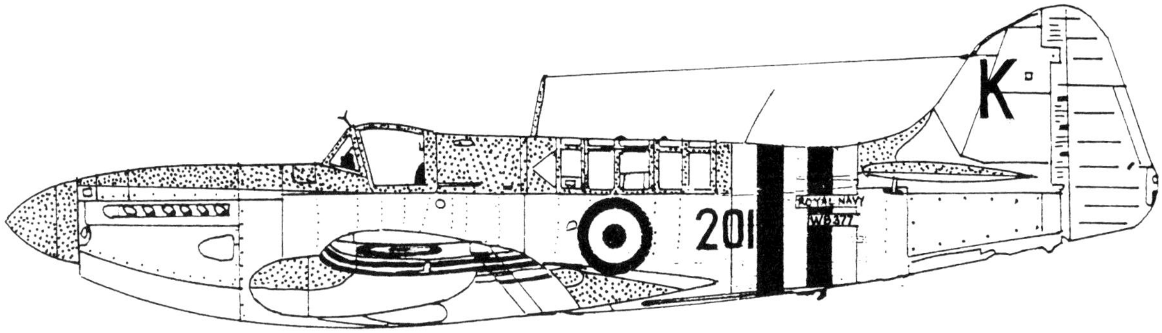

Firefly WB377/201K of 817 Sqn aboard HMAS *Sydney*, Korea, 1951
Standard scheme of extra dark sea grey upper surface, including spinner, with sky fuselage and undersurface. Black numbers, serial and code. Black and white stripes encircle fuselage and wings.

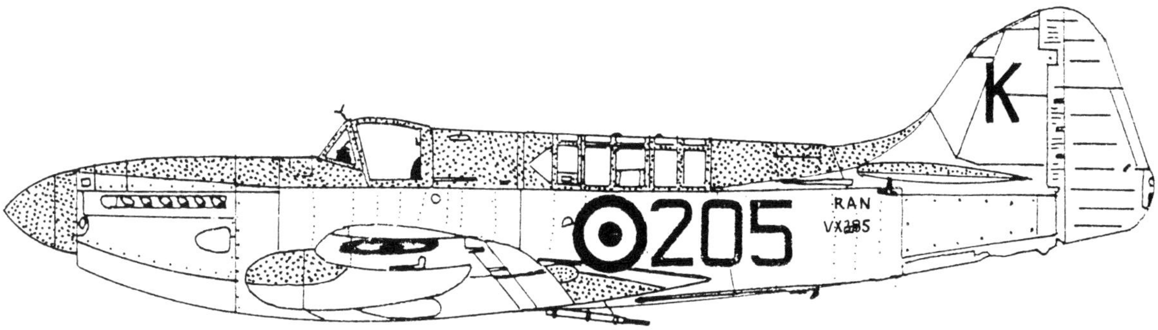

Firefly VX385/205K of 717 Sqn aboard HMAS *Sydney*, 1953
Standard scheme.

Firefly WD826/891 target tug
Standard scheme. Black stripes on undersurface. Black and yellow spinner. Yellow band encircles fuselage, although not over black sections of the undersurface. Yellow bands on wing upper surface.

Firefly WB518/842NW target tug of 851 Sqn
Silver overall, including wing tanks. Black and yellow spinner. Black numbers, serial and lettering. Yellow band on fuselage side and wings upper surface. Yellow and black striped undersurface.

Firefly VX373 trainer conversion in delivery markings, 1953
Silver overall. Black serial. Yellow bands encircle fuselage and wings.

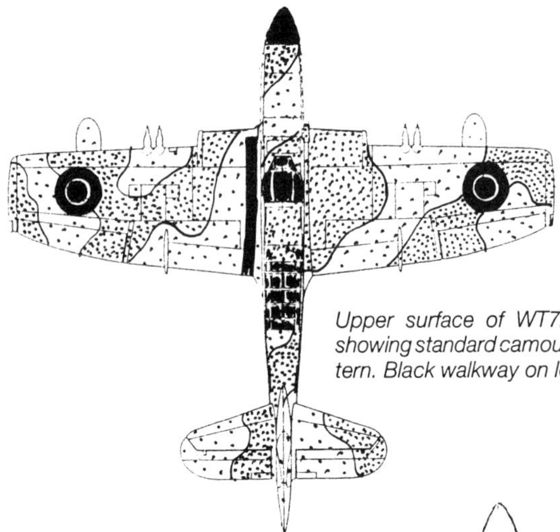

Upper surface of WT726/228K, showing standard camouflage pattern. Black walkway on left wing.

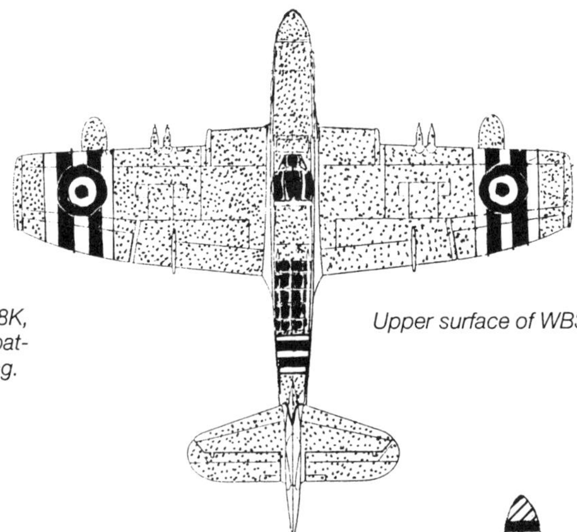

Upper surface of WB377/201K.

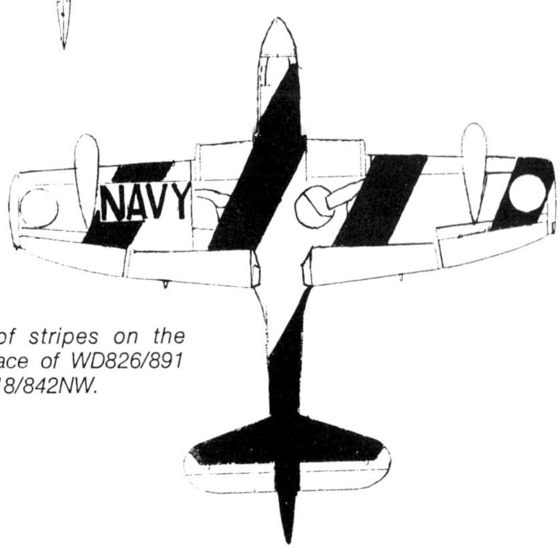

Pattern of stripes on the undersurface of WD826/891 and WB518/842NW.

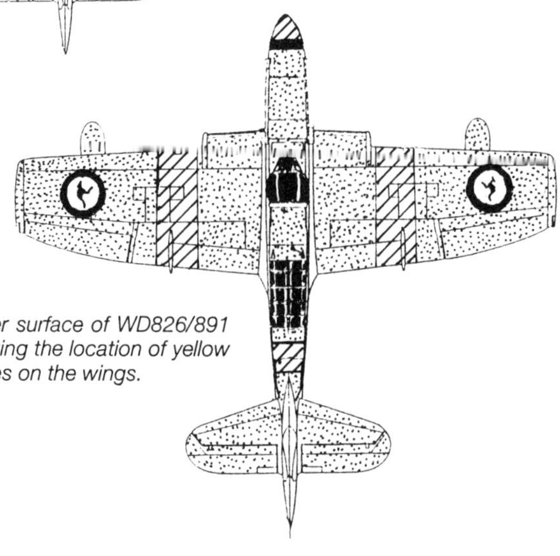

Upper surface of WD826/891 showing the location of yellow stripes on the wings.

DE HAVILLAND
SEA VENOM

Royal Australian Navy de Havilland Sea Venom WZ896 of 808 Squadron, the first RAN unit to equip with the all weather fighter. (RAN)

DE HAVILLAND SEA VENOM

The second of de Havilland's trio of twin boom jet fighters which began with the Vampire and ended with the Sea Vixen, the DH.112 Venom was developed as an interim replacement for the Royal Air Force's Vampire fighter-bombers as a result of the delayed introduction to service of the Hawker Hunter and Supermarine Swift.

The Swift proved to be unsatisfactory and served in only small numbers, while the Hunter was used only as an interceptor until later in its career. The result was that the Venom was used for a longer period than had been originally planned, filling the RAF's ground attack needs until the Hunter FGA.9 came along.

The Vampire had been an extremely successful design, built in single seat fighter-bomber and two seat night fighter and trainer variants for the RAF and more than two dozen other nations, including Australia. Licence production was undertaken in Australia, France, Italy, India and Switzerland, the overall production tally reaching 3,987 aircraft between the first flight of the prototype in September 1943 and the closure of the line in 1961.

While it is true to say the Venom was developed from the Vampire, it was nevertheless a new design in many ways. The Sea Vixen was a very different animal again, sharing only the twin boom configuration with its predecessors.

Although nowhere near as well known as the Vampire, the Venom, too, was a sales success, a total of 1,483 of all versions being built, of which 393 were navalised Sea Venoms. Venom operators were Britain, Italy, Iraq, Venezuela, Sweden and Switzerland, while Sea Venoms were purchased by Britain, Australia and France. Australia operated 39 examples and France 94, most of the latter manufactured under licence as the Aquilon.

The biggest export operator of the Venom was Switzerland, which flew 250 locally built single seaters from 1953 and retired the last examples as late as 1983.

For Australia's Fleet Air Arm, the Sea Venom represented half of the second 'pair' of fixed wing combat aircraft to fly from the Royal Australian Navy's carriers, beginning with the Sea Fury and Firefly, progressing through the Sea Venom and Gannet and ending with the Skyhawk and Tracker.

From Vampire to Venom

Even before the Vampire entered operational service with the Royal Air Force in 1946, de Havilland had begun planning better performing versions powered by the new de Havilland Ghost centrifugal flow turbojet which promised considerably greater power than the Goblin engine in the Vampire. Most Vampires were powered by the Goblin which offered between 3,100 and 3,500lb thrust, although some (Australia's Mk.30 and France's Mk.51 and *Mistrel*) had larger 5,000lb thrust Rolls-Royce Nenes installed.

The Ghost produced 4,850lb thrust in its earlier production versions and 5,300lb later on, and its similar dimensions to the Goblin made it an obvious candidate for installation in the Vampire. The third production Vampire (TG278) was therefore fitted with a Ghost engine and made its first flight with this powerplant on 8 May 1947 from de Havilland's facility at Hatfield on the northern outskirts of London. The Ghost was rated at 4,400lb thrust at this stage and the combination proved successful, test pilot John Cunningham taking the Ghost-Vampire to a world altitude record of 59,446 feet (18,131m) in March 1948 after the fitting of a pressure cabin and wings of greater span.

It was realised at an early stage that to take full advantage of the extra power offered by the Ghost, de Havilland would have to develop a new, 'thin' wing for the aircraft. Under the designation DH.112, work proceeded along these lines, initially as a private venture.

Official recognition of the project came in 1949 when the new aircraft (to that point

Restored Sea Venom WZ895 with canopy open and wings folded. The two crew members sat almost side-by-side for better communication, as in the Mosquito. (via Steve Long)

The first of de Havilland's twin boom jet fighters, the DH.100 Vampire. (DoD)

The Vampire was developed into two seat versions including the DH.115 Vampire Trainer. (DoD)

(left) A link in the chain which connected the Vampire to the Venom and Sea Venom was the DH.113 Vampire night fighter with two seats and airborne interception radar.

known internally as the Vampire Mk.8) was made the subject of Air Ministry Specification F.15/49 and named 'Venom'.

Although very similar to the Vampire, there were several notable differences apart from the installation of the more powerful engine. The new thinner wing was of slightly greater area and revised aerofoil section and featured a modest 17.1 degree sweepback. A single stores hardpoint capable of carrying up to 1,000lb (454kg) was fitted outboard of each inward retracting main undercarriage leg and wing tip tanks of 75imp gal (341 l) capacity were fitted. Although removable on the ground, these tanks were not jettisonable in flight and were designed to be retained in all flight regimes including combat.

The greater volume of air required by the single Ghost engine led to a redesign of the wing root air intakes, while some subtle reshaping of the fuselage aft of the cockpit and a larger tailpipe were also necessary. Most other systems and components were refined to some extent, but the major structural features of the Vampire were retained, among them the fabric covered laminated balsa-plywood sandwich construction of the fuselage pod, a legacy of the Mosquito. Otherwise, the Venom's structure was metal.

The first of two Venom prototypes – VV612 – was flown from Hatfield by John Derry on 2 September 1949, followed by VV613 in July 1950. The first aircraft was not initially fitted with the wingtip tanks. Despite it being designed primarily as a ground attack aircraft, the Venom soon proved itself as a fighter, tests against the RAF's standard day fighter of the time, the Meteor F.8, highlighting the de Havilland aircraft's higher speed, greater rate of climb, higher ceiling, greater manoeuvrability and steadier platform for its four 20mm Hispano cannon mounted in the lower nose beneath the cockpit floor.

The Venom offered considerably enhanced performance over its predecessor, its maximum speed of 640mph (1,030km/h) being some 90mph (145km/h) faster than the Vampire FB.9 and its 9000 feet (2743m) per minute initial climb rate was nearly twice that of the Vampire. The Vampire FB.9 had a maximum weight of 12,390lb (5,620kg) while the Venom weighed in at 15,400lb (6,985kg) loaded.

The Venom was not without fault in its early days, however. Its rate of roll was described as being 'poor' without the tip tanks fitted and even worse with them, while the wings had to be fitted with drooped leading edges to cure a tendency to pitch up at medium and high altitudes. The first Venoms were not fitted with ejection seats, an omission which was rectified later.

Powered by Ghost 103 engines of 4,850lb thrust, Venom FB.1 deliveries to the Royal Air Force began in April 1952, after which production moved from Hatfield to Chester, while some aircraft were assembled by Fairey Aviation at Ringway (Manchester) and by Marshalls of Cambridge. A total of 375 Venom FB.1s was manufactured until 1954 when production switched to the next single seat variant, the FB.4.

This Venom (151 of which were built) featured several improvements including revised rudder shapes to prevent excess yaw and possible rudder locking at low speeds, hydraulically operated ailerons and the ability to carry underwing drop tanks.

Like the Venom FB.1, the Mk.4 proved to be a very successful ground attack aircraft and most squadrons were based overseas rather than in the United Kingdom. Deployments included West Germany and the Middle East, the latter seeing the Venom in action against forces on the ground during the Suez crisis of 1956 and in Yemen the following year. RAF Venom squadrons also took part in operations against communist terrorists in Malaya, as did pilots of

The DH.112 Venom prototype (VV612) displays its new thinner and slightly swept back wing plus its tip tanks. These and the installation of a more powerful Ghost engine were the most obvious differences between the early Venoms and the Vampire.

(left) The prototype Venom NF.2 night fighter, predecessor of the Sea Venom. The prototype first flew in August 1950 without tip tanks and carrying the Class B civil registration G-5-3.

No 14 Squadron RNZAF, which flew Venoms hired to the New Zealand government by the RAF.

Night Fighters

The direct predecessor of the Sea Venom was the land based Venom NF.2 night fighter, a two seater developed in a similar manner to the previous Vampire night fighter.

This variant required a new longer and wider forward fuselage to accommodate the pilot and the radar operator/navigator, who sat to the right of and slightly behind the pilot, and the nose mounted AI Mk.10 airborne interception radar. Construction of the fuselage pod remained as before – plywood balsa sandwich.

The crew sat under a heavily framed upward hinging canopy, although this was replaced later by a clear view canopy which was jettisonable. Ejection seats were not fitted to any night fighter Venoms, a point which attracted considerable criticism.

Taller, pointed fins also distinguished this new family of Venoms, the first of which flew from Hatfield on 22 August 1950 in the hands of John Derry. Developed as a private venture, it flew in British Class B civil markings as G-5-3 initially before being adopted by the RAF and given the serial number WP227. Early flights were conducted without the wingtip tanks attached. Like its single seat counterparts, the night fighter Venom was criticised for its poor rate of role and excessive yaw, characteristics which deemed it to be marginal as a night fighter.

Various remedies were attempted, including the replacement of the original 'bullet' fin fairings with an elongated fillet and the fitting of more responsive ailerons.

Powered by 4,850lb thrust Ghost 103s, the first of 90 production Venom NF.2s was flown in March 1952 and deliveries to the RAF's No 23 Squadron began six months later. As it turned out, this squadron was the only operational unit to fly the NF.2,

A production Venom NF.3, which apart from changes around the jet pipe, looks very much like a Sea Venom.

(right) A Swiss Venom FB.4, built under licence. Switzerland was a large user of the Venom and retired its last examples as late as 1983 after three decades' service.

DE HAVILLAND DH.112 SEA VENOM FAW.53
Powerplant: *One de Havilland Ghost Mk.104 centrifugal flow turbojet engine rated at 4,950lb thrust at sea level. Engine comprises single stage centrifugal compressor and single stage axial-flow turbine, weight 4,950lb (2,245kg). Fuel capacity 456imp gal (2073 l) comprising 216imp gal (982 l) in eight wing tanks, 90imp gal (409 l) in one mid fuselage tank and 150imp gal (682 l) in two non jettisonable wingtip tanks.*
Dimensions: *Wing span 42ft 10in (13.05m); length 36ft 7.25in (11.15m); height (wings extended) 8ft 6in (2.59m), height (wings folded) 9ft 7in (2.92m); width (wings folded) 23ft 0in (7.01m); wheel track 11ft 4in (3.45m); wing area 279.75 sq ft (25.99sq m).*
Structure: *Balsa-plywood sandwich fuselage nacelle with separate nose radome, engine cowling and jet pipe assemblies. Wings of metal cantilever two spar construction, twin light alloy tail booms with single fin and rudder assembly at aft of each, light alloy tailplane and elevator assembly. Hydraulically powered ailerons. Hydraulically operated retractable tricycle undercarriage with Maxaret anti-skid units fitted to the outwards retracting main wheels. Fully castering, self centering rearward retracting nosewheel. Jettisonable (by cartridge) rear hinging cockpit canopy.*
Systems and Equipment: *Cabin pressurisation (to 3psi) and air conditioning; camera gun under port inboard wing; two Martin Baker Mk.4A ejection seats; dual 28 volt DC electrical systems; AI Mk.17 radar in nose radome; radar altimeter, IFF, UHF radio, VHF radio, intercom; dual hydraulic system for undercarriage, flaps, brakes, dive brakes, arrestor hook, windscreen wiper, wing folding mechanism and powered ailerons; pneumatic system for canopy seal and RATOG (when fitted).*
Armament: *Four 20mm Hispano Mk.V cannon mounted in lower fuselage under cockpit with 150rpg; provision for eight 60lb (27kg) rocket projectiles under wings.*
Weights: *Max takeoff weight (external armament) 15,900lb (7,212kg); max landing weight (arrested) 12,220lb (5,543kg); normal landing weight (runway) 13,200lb (5,987kg); empty approx 11,000lb (4,990kg).*
Performance: *Max speed 563mph (906km/h) at sea level, 558mph (898km/h) at 20,000ft (6,100m), 544mph (876km/h) at 30,000ft (9,150m); patrol speed 339mph (546km/h) at 30,000ft (9,150m); rate of climb (max weight) 4850ft (1478m)/min at sea level, 3,000ft (914m)/min at 20,000ft (6,100m), 1,900ft (579m)/min at 30,000ft (9,150m); time to 20,000ft (6,100m) 6.1min, to 30,000ft (9,150m) 10.2min, to 40,000ft (12,192m) 18.6min; service ceiling 40,000ft (12,192m); patrol duration 2.0hrs at 30,000ft; max range 1,000 miles (1609km).*

mainly due to the discovery of various defects, not the least of which was a structural weakness in the wings which was thought to be the cause of several crashes.

The high accident rate during night approaches resulted in the fitting of the clear view canopy and other modifications were made to the fin and rudder, resulting in the definitive shape as fitted to the Sea Venom. Modified aircraft were designated the Venom NF.2A and these served with two RAF squadrons.

The Venom NF.3 was a new build version incorporating the NF.2A's modifications along with improved radar, powered ailerons and a new tailplane design which was housed wholly between the fins with no external extension.

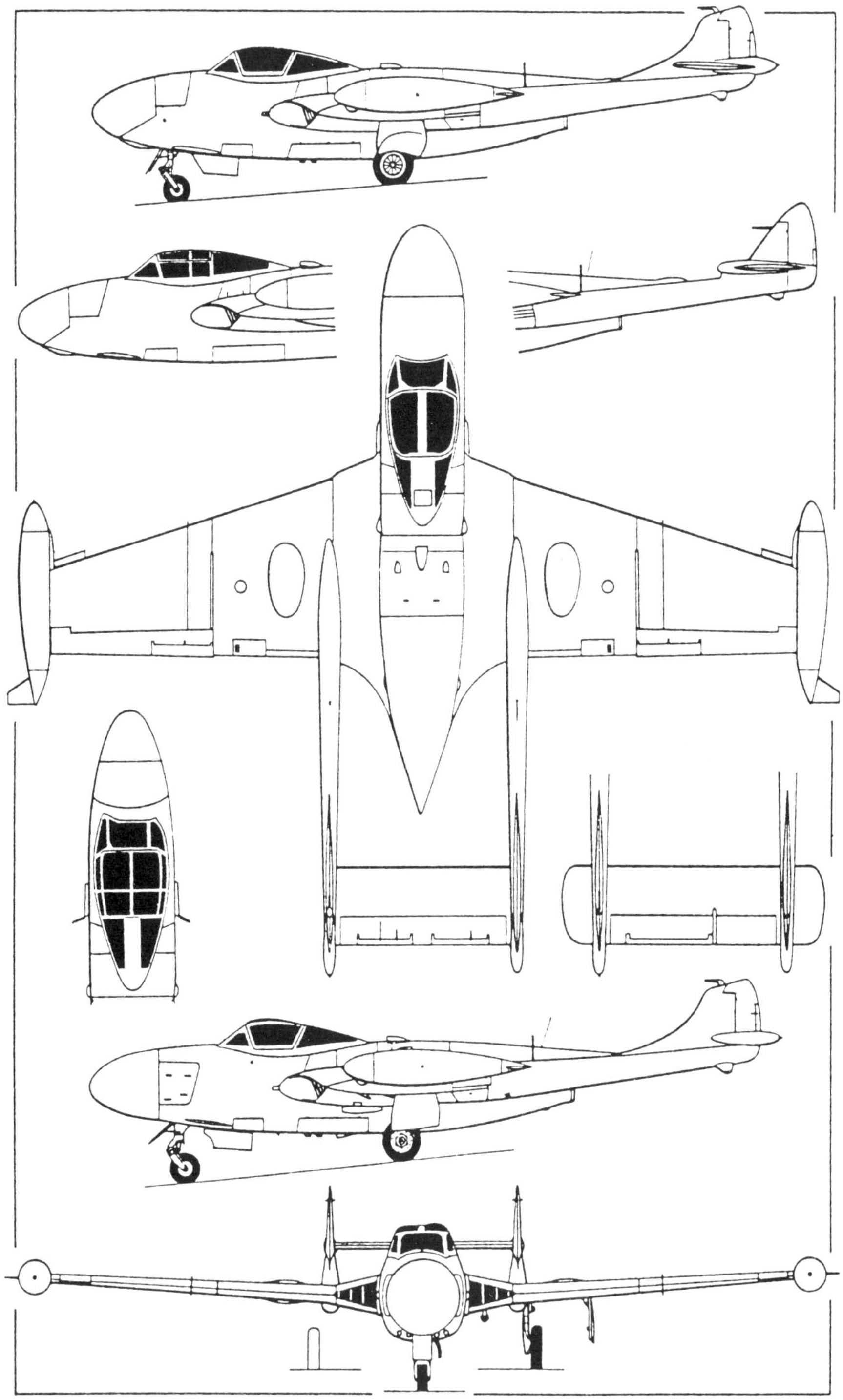

The main three view shows the Sea Venom FAW.22 and is also applicable to Australia's FAW.53. The upper side view is of the FAW.21 while the scrap views show the FAW.20 (with old style canopy) and the early fin and rudder, tailplane and canopy designs.

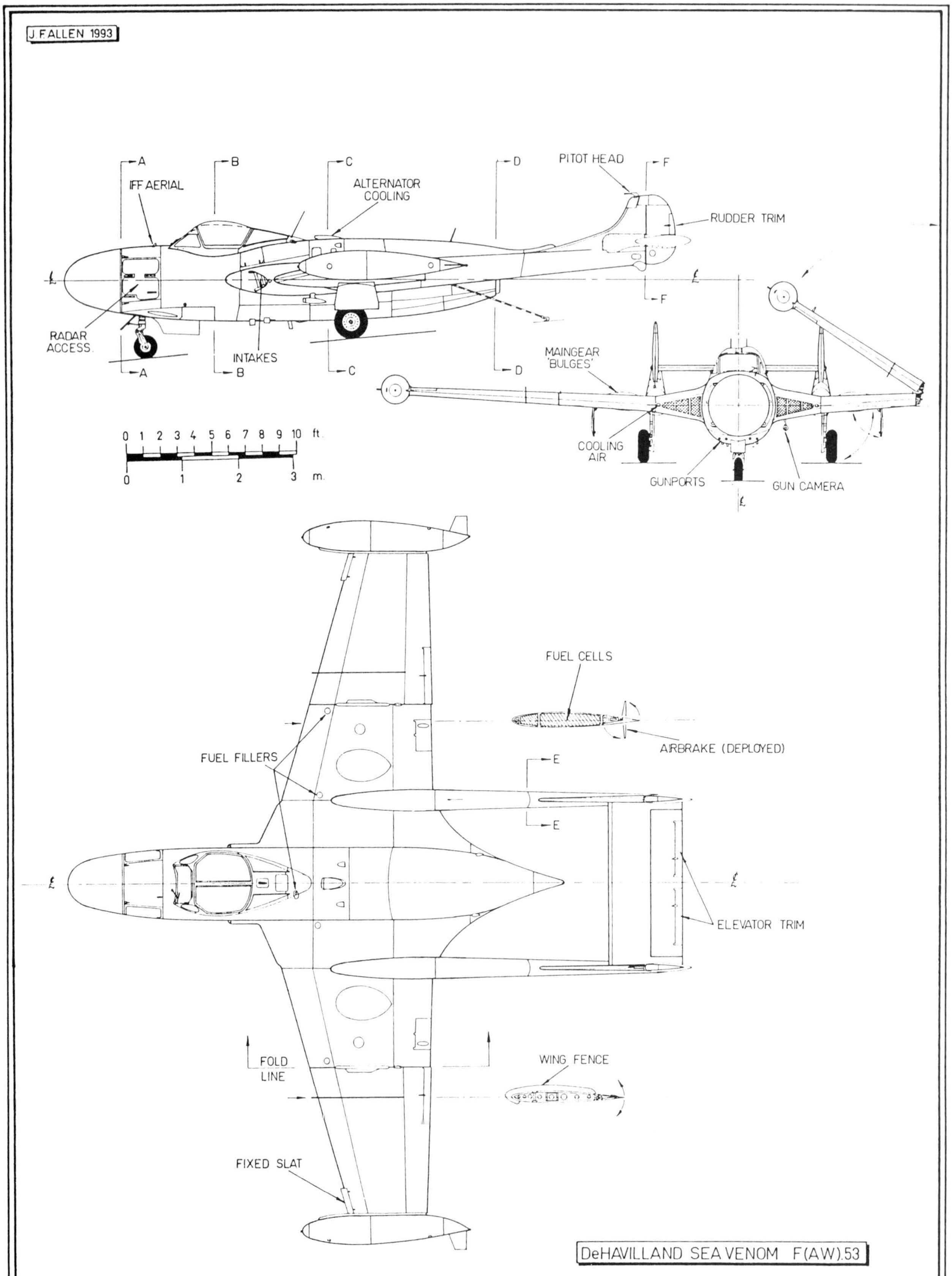
J.F.ALLEN 1993
A
B
C
D
F
E
IFF AERIAL
ALTERNATOR COOLING
PITOT HEAD
RUDDER TRIM
RADAR ACCESS
INTAKES
MAINGEAR 'BULGES'
COOLING AIR
GUNPORTS
GUN CAMERA
0 1 2 3 4 5 6 7 8 9 10 ft.
0 1 2 3 m.
FUEL CELLS
AIRBRAKE (DEPLOYED)
FUEL FILLERS
ELEVATOR TRIM
FOLD LINE
WING FENCE
FIXED SLAT
DeHAVILLAND SEA VENOM F(AW).53

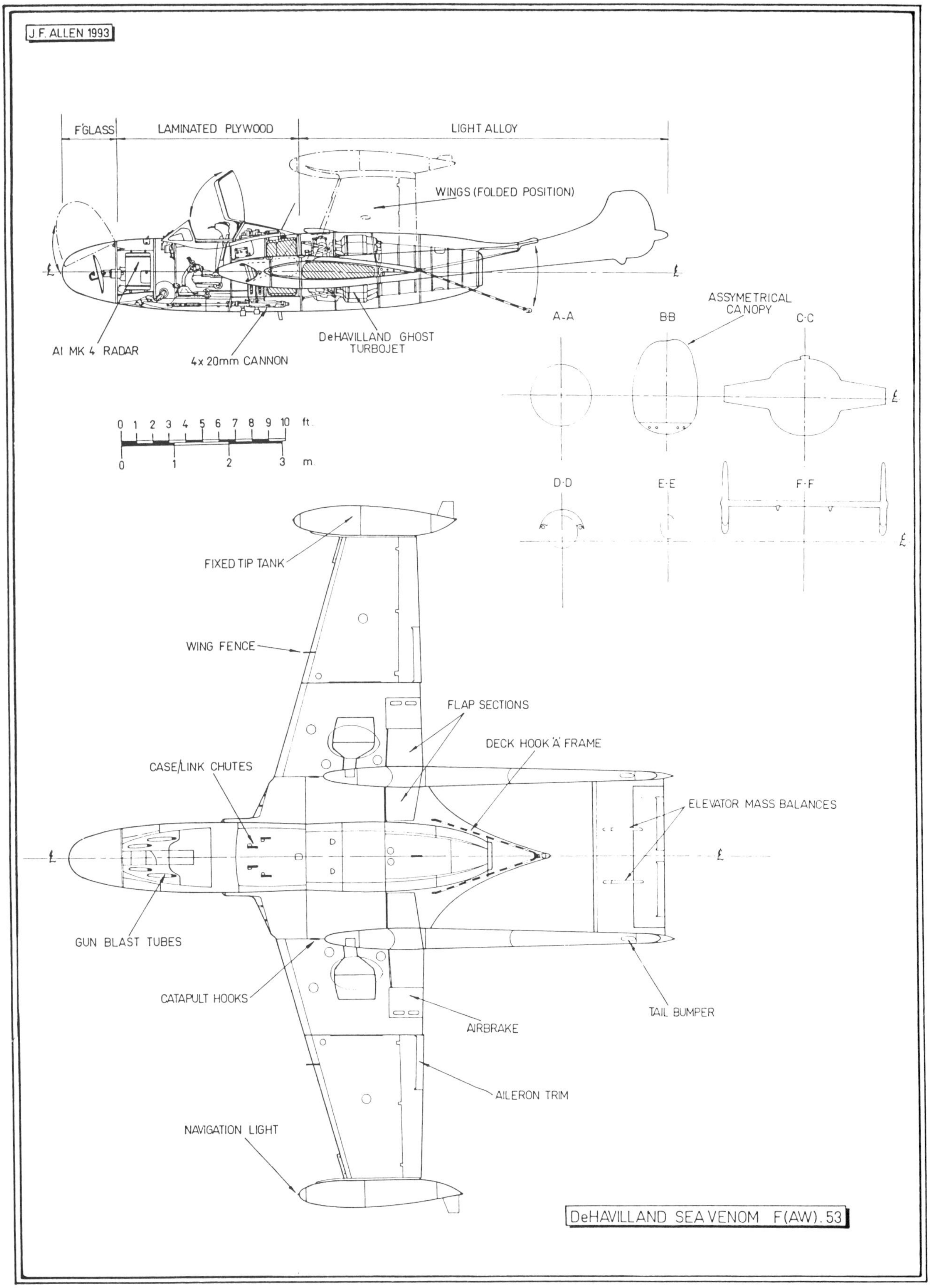
J.F. ALLEN 1993
F'GLASS
LAMINATED PLYWOOD
LIGHT ALLOY
WINGS (FOLDED POSITION)
AI MK 4 RADAR
4x 20mm CANNON
DeHAVILLAND GHOST TURBOJET
ASSYMETRICAL CANOPY
A-A
BB
C-C
D-D
E-E
F-F
0 1 2 3 4 5 6 7 8 9 10 ft.
0 1 2 3 m.
FIXED TIP TANK
WING FENCE
FLAP SECTIONS
DECK HOOK 'A' FRAME
CASE/LINK CHUTES
ELEVATOR MASS BALANCES
GUN BLAST TUBES
CATAPULT HOOKS
TAIL BUMPER
AIRBRAKE
AILERON TRIM
NAVIGATION LIGHT
DeHAVILLAND SEA VENOM F(AW). 53

The Sea Venom prototype WK376 was first flown in April 1951 and featured a different canopy, vertical tail surface and tailplane design to that fitted to later models. At this stage, ejection seats were not fitted.

The first Venom NF.3 flew in February 1953 and a total of 129 was built, serving with five RAF squadrons until replaced by Gloster Javelins from 1957. Only one Venom night fighter export sale was recorded, 62 NF.51s being delivered to Sweden between 1952 and 1957. Early aircraft were equivalent to the RAF's NF.2 and later examples were to NF.3 standards.

The Hooked Venom

Naval operations of de Havilland's twin boom fighters were not a new phenomenon by the time development of a carrier based derivative of the Venom NF.2 began in 1950. In December 1945, Lt Cdr Eric ('Winkle') Brown had recorded the world's first jet aircraft deck landing when the suitably modified second prototype Vampire arrived aboard HMS *Ocean*.

With an arrestor hook, larger flaps and longer travel undercarriage legs, this aircraft was redesignated Sea Vampire Mk.10 and subsequent trials resulted in the placing of an order for a small number of Sea Vampires for the Royal Navy.

These were not intended for regular carrier use but as shore based trainers giving RN pilots experience in jet aircraft. However, operations from carriers by these aircraft did build up a bank of experience which would prove useful when the Sea Venom came along.

Six Sea Vampire F.21s (modified from land based Mk.3s) were used in the extraordinary 'undercarriageless deck landing trials' of 1949, when a flexible flight deck material made of rubber was laid down and the aircraft landed wheels up after a conventional catapulted departure from the carrier.

The idea was to save weight and complexity by removing the aircraft's undercarriage, and although 'Winkle' Brown made 40 such landings at Farnborough and aboard HMS *Warrior*, the idea came to nought. Perhaps someone realised that from time to time even naval aircraft need to operate from conventional runways!

The Sea Venom came about following Fleet Air Arm evaluation of the Venom NF.2 and via Specification N.107 for a carrier-based all weather fighter, an interim design which could fill the gap between the piston engined de Havilland Sea Hornet and the much more advanced Sea Vixen which was still some years away.

Design of what would be the Royal Navy's first two seat all weather jet fighter was undertaken at the old Airspeed factory at Christchurch rather than at the somewhat overstretched Hatfield, under the leadership of W A Tamblin.

The obvious changes which transformed a Venom into a Sea Venom were power operated upward folding wings (the fold occurring at about mid span) and the fitting of catapult attachment points and arrestor hook. The latter was a V-frame design which was stowed in an extended fairing above the jet pipe. The undercarriage was strengthened to cope with the Sea Venom's heavier weights and the stresses involved with deck landings and from the second production model, a longer stroke undercarriage was fitted to further assist in these areas.

Otherwise, the Sea Venom as originally presented was pretty much as per its land based counterpart with the Venom NF.2's salient features including the heavily framed canopy, small vertical tail surfaces, tailplane extending past the fins, a 4,850lb thrust Ghost 103 engine, and manual ailerons.

Three prototypes were ordered, the first two of which (WK376 and WK379) were built in Hatfield's Experimental Department and the third (WK385) at Christchurch. Production would be shared between Christchurch and Chester.

WK376 recorded its maiden flight on 19 April 1951 (eight months after the first Venom night fighter had flown) followed by the other two prototypes of which WK385 was the last to fly in July 1952. This aircraft was the first to be fitted with folding wings. Deck landing trials began in July 1951 aboard HMS *Illustrious* using the first prototype.

Assessment of the aircraft was generally favourable, the Aircraft and Armament Experimental Establishment (A&AEE) at Boscombe Down generally praising its deck takeoff and landing characteristics, but criticising its arrestor hook damping qualities and its lateral and directional control characteristics on landing using 'standard techniques'. Boscombe Down considered that the use of the newly developed mirror deck landing aid would help.

Into Production

Production of the Sea Venom FAW.20 began in 1953, the first production aircraft (WM500) flying in March of that year. The first FAA squadron to re-equip with the Sea Venom was No 890, which swapped its Supermarine Attackers for the new aircraft in March 1954.

Ultimately, seven Royal Navy Fleet Air Arm squadrons operated Sea Venoms, flying from the carriers *Albion*, *Ark Royal*, *Eagle*, *Bulwark*, *Centaur* and *Victorious*. Most front line squadrons had replaced their Sea Venoms with Sea Vixens by 1958-59 or had been disbanded, but one unit (894 Squadron) used its Venoms until 1960.

Sea Venom FAW.20s featured the redesigned fin, rudder and fin fairings which had been fitted to later night fighter Venoms, non jettisonable tip tanks and folding wings and a cockpit canopy which was capable of being jettisoned underwater. For the moment, ejection seats were not fitted.

Operational equipment included Airborne Interception Radar, four 20mm Hispano Mk.V cannon with 150 rounds per gun and provision to carry eight rocket projectiles or two 500lb (227kg) bombs under the wings.

Total fuel capacity was 456imp gal (2,073 l) divided between fuselage, wing and wingtip tanks; the cabin was pressurised to 3psi; the hydraulic system operated the undercarriage, flaps, brakes, dive brakes, arrestor hook, powered ailerons (on later models) and wing folding mechanism. The canopy seal was pneumatically operated.

Fifty production Sea Venom FAW.20s were built but only a year after the type entered service it was experiencing difficulties which led to its withdrawal from first line squadron service.

The most serious of these was a tendency for the arrestor hook to break when it picked up a wire, the result usually being the aircraft careering off the flight deck and depositing itself in the sea as there was insufficient time to spool up the Ghost and attempt a go-around. This made life difficult for the crews, who on more than one occasion had to wait in their submerged aircraft while the ship passed over them. Remember, there were no ejection seats in early Sea Venoms!

Production Improvements

Many of the problems associated with the Sea Venom FAW.20 were solved in the second production model, the FAW.21. First flown in May 1954, a total of 167 was

Front and rear views of the Sea Venom showing the wings in the process of folding (top) and fully folded. The rear view shows some detail of the 'pen nib' fairing over the jet pipe for the arrestor hook. (via Steve Long/RAN)

built and improvements included the fitting of powered ailerons, longer stroke undercarriage, clear view canopy, Martin Baker Mk.4 ejection seats for both crew members, a strengthened arrestor hook and Maxaret anti-skid brakes.

Provision was made for the installation of rocket assisted takeoff gear (RATOG) and power was initially provided by a 4,950lb thrust Ghost 104, later upgraded to a 5,300lb thrust Ghost 105 as fitted to the final production model, the Sea Venom FAW.22.

This last Sea Venom variant for the Royal Navy was similar to late model FAW.21s with Ghost 105 engines and the aerodynamic improvements (such as the new horizontal tail surfaces design) which had been incorporated in later land based Venoms. Production of the FAW.22 began in 1956 and 39 were had been built by the time the last was delivered in January 1958.

Royal Navy Sea Venoms from HMShips *Albion* and *Eagle* saw active service during the Suez crisis of late 1956, mounting attacks on Egyptian military convoys and airfields in support of friendly forces on the ground. During these actions the Sea Venoms destroyed several MiG-15s on their airfields.

The only export orders received for Sea Venoms were from the Royal Australian Navy and France's *Aeronavale*.

Australia's aircraft were based on the FAW.21 and designated the Sea Venom FAW.53 with special equipment for the RAN's needs. Thirty-nine were delivered in 1955-56 to fly from HMAS *Melbourne* and as described in more detail in the following chapters, these aircraft remained in first line service until replaced by A-4 Skyhawks in 1967.

France's 94 Sea Venoms were built under licence by Sud Aviation and given the name *Aquilon* ('North Wind') for operation from the carrier *Clemenceau*. Aquilons were built in three production versions and changes introduced included the fitting of ejection seats from the start, strengthened undercarriage and a rearward sliding canopy in place of the standard hinged unit.

An interesting variant unique to the French was the Aquilon 203, a single seater equipped with American APQ-94 radar and a new canopy design. Forty of these were built. The first *Aeronavale* Aquilon was flown in October 1952 and the type withdrawn from carrier service in 1962.

A preserved former RAN Sea Venom, WZ939 (via Steve Long)

AUSTRALIA'S SEA VENOMS

By the middle of 1951, Australia's young Fleet Air Arm was busy. The needs of the Korean war would shortly see the aircraft carrier HMAS *Sydney* depart for those troubled waters to have her Sea Furies and Fireflies fly more than three months of intensive operations against the communists.

The absence of *Sydney* and the delayed commissioning of the fleet's second carrier, HMAS *Melbourne*, resulted in the loan of HMS *Vengeance* to the Royal Australian Navy from late 1952 until *Melbourne* was commissioned three years later.

At the same time, the future aircraft equipment needs of the FAA were being examined. The piston engined Hawker Sea Fury and Fairey Firefly were in themselves fine aircraft, but the jet age was upon us, and it was in this area that options for replacements were examined.

As had been the case with the piston engined aircraft, commonality with the Royal Navy was sought, with the result that the choice came down to either the Hawker Sea Hawk or de Havilland Sea Venom replacing the Sea Fury in the fighter role, and the Fairey Gannet turboprop replacing the Firefly in the anti submarine role.

The decision between the single seat Sea Hawk and two seat Sea Venom was that the former, although offering similar performance at a lower cost was very much a day fighter, lacking radar. The Sea Venom, on the other hand, was designed from the start as an all weather fighter and had a secondary (albeit limited) ground attack capability and consequently better met Australia's needs. Unlike Britain, Australia could not afford the luxury of two types to satisfy its overall requirements.

The reasons for the RAN's choice of the Sea Venom were noted as being its ability to seek and destroy enemy 'pro submarine' aircraft in any weather, day or night; the reduced performance of an all weather fighter such as the Sea Venom (when compared to land based fighters) would not be a major factor as it would be more likely to meet enemy long range aircraft rather than fighters; and its ground attack capability with rocket projectiles.

Consequently, on 23 July 1951, the Minister for the Navy, Mr William McMahon, announced the de Havilland Sea Venom would be ordered for the Royal Australian Navy. The order was made formal in December of the same year when a contract was placed for 39 aircraft by the British Air Ministry on behalf of the RAN. The Fairey Gannet was chosen as the RAN's new anti submarine aircraft the following year, while eight spare Ghost engines were also ordered for the Sea Venoms.

Licence Production Discussed

The decision to operate the Sea Venom and Gannet from the RAN's carriers meant they would have to be modified to accept the faster and heavier aircraft. Initial discussions centred around modifying both HMAS *Sydney* and the still incomplete HMAS *Melbourne* to incorporate an angled flight deck, steam catapult and mirror deck landing system, but in the event only *Melbourne* was so modified. At about the time of the orders for the new aircraft, it was announced that *Sydney* would be relegated to a non flying training role when *Melbourne* entered service.

The substantial modification work further delayed *Melbourne's* entry into service, but at least brought her up to the latest operational standards and ensured she would remain viable for many years ahead. The ship and her new aircraft entered service simultaneously.

In April 1953 the de Havilland company's Australian subsidiary, De Havilland Ltd (later Hawker de Havilland) approached the Australian government about building the Sea Venom at its Bankstown NSW facility.

The company had built 212 Mosquitos there in the last years of World War II and subsequently, and at the time was nearing completion of its Vampire manufacturing programme for the Royal Australian Air Force. The company was keen to provide continuity of work for itself and its employees and built 80 single seat Vampires from 1949 and 110 Vampire Trainers subsequent to those. The largest batch of the latter was ordered after the Sea Venoms were contracted for, and plans to manufacture the naval fighter were dropped.

Australia's Sea Venoms

The 39 Sea Venoms for the Royal Australian Navy were designated Sea Venom FAW.53s, the appellation given to export models of the aircraft. The FAW.53 was based on the Royal Navy's FAW.21 with

One of several RAN Sea Venoms which survive today, WZ931 is prepared for flight in its operational days under a threatening sky at Nowra. (RAN)

A neat quartet of Sea Venoms from 808 Squadron RAN. The 'Y' identification letter on the fin was used by 808 when it operated from HMAS Melbourne. *(RAN)*

4,950lb thrust Ghost 104 engine, powered ailerons, long stroke undercarriage and provision for Martin-Baker Mk.4A ejection seats. They differed mainly in the radar equipment fitted, this being the British Airborne Intercept (AI) Mk.17 type. Ejection seats were not initially fitted, this modification being incorporated by Hawker de Havilland in Australia in early 1957.

Australia's Sea Venoms came from the fourth production batch of a planned 49 FAW.21/53s ordered under contract 6/Aircraft/7561/CB.7 (a) placed in January 1952. All aircraft were built at de Havilland's Christchurch (near Bournemouth) facility but amendments to the contract resulted in the last ten aircraft being cancelled and the remaining 39 all being built as FAW.53s for the Royal Australian Navy.

The aircrafts' serial numbers were WZ893-911 and WZ927-946. These were the 750th to 788th of the DH.112 Venom series to be manufactured, their constructor's numbers (12750-12788) reflecting that.

In Australian service, the Sea Venoms carried the British serial numbers for most of their careers, although new serial prefixes were applied to RAN aircraft generally from 1965. The result was that some of the Sea Venoms which survived after that date appeared with their allocated 'N4' prefix followed by the three digits of the existing serial number. For example, WZ935 became N4-935.

The Sea Venom FAW.53s were initially delivered to the Royal Navy between February 1955 and January 1956 before being transferred to the RAN. As the prototype FAW.53, WZ893 spent some time with the manufacturer for trials and development work and at Boscombe Down for evaluation. WZ894, 895 and 955 also spent brief periods of time at Christchurch performing development flying duties.

Just off! 808 Squadron Sea Venom WZ897 at the moment of departure from HMAS Melbourne. *(via Steve Long)*

Sea Venoms Delivered

While her new Sea Venoms (and Gannets) were being prepared for handover to the RAN, the ship which would shortly become the service's flagship, HMAS *Melbourne* was also being readied. The aircraft carrier was commissioned into service on 28 October 1955 at Barrow-in-Furness, Lancashire, in a ceremony performed by Lady White, the wife of the Australian High Commissioner in the United Kingdom, Sir Thomas White.

The ship then carried out working up trials which were successfully completed on Australia Day (26 January) 1956 when Sea Venoms and Gannets were launched and recovered.

In the meantime, the RAN Sea Venom crews had been undergoing training on their aircraft by 'X' Flight of the Royal Navy's 891 Squadron at NAS Yeovilton in Somerset. 891 Squadron was equipped with Sea Venom FAW.21s or FAW.20s upgraded to the later standard.

Smoke on go! The 'Ramjets' display team formed by 724 Squadron in 1959. (via Steve Long)

(right) 805 Squadron's WZ929 aboard HMAS Melbourne, *with 'M' identification letter and chequerboard tip tank pattern clearly seen.*

Sea Venom WZ939 of 'B' Flight, 816 Squadron on the ground at Nowra. 816 Squadron was the last unit to fly Sea Venoms from HMAS Melbourne, *in company with its Fairey Gannets. (via Steve Long)*

808 Squadron's WZ896 in flight. When it converted to the Sea Venom it became the first Commonwealth naval aviation squadron to be equipped with an all weather jet fighter. (RAN)

An 808 Squadron Sea Venom aboard HMAS Melbourne *during the squadron's three years' flying from the flagship. (RAN)*

Training of the Australian crews had begun in March 1955 and by the following August the time had come to commission the first Royal Australian Navy Sea Venom squadron. As such 808 Squadron RAN became the first Commonwealth naval aviation unit to be equipped with an all weather jet fighter.

In November 1955, 808 Squadron had eight Sea Venoms on strength (WZ894-WZ901) and after a further working up period, was embarked aboard *Melbourne* in March 1956. As noted above, WZ893 was also regularly flying during 1955 and came onto the squadron's strength in the same month. The other 30 Sea Venoms of the batch were delivered during 1955 and early 1956 to the Royal Naval Aircraft Yard at Abbotsinch where they were stored in protective cocoons and crated for shipment. They were then transported to Glasgow, from where *Melbourne* would take them on board for the journey to Australia.

The Sea Venoms were formally taken on RAN charge on 27 February 1956, while *Melbourne* sailed from Glasgow on 11 March bound for Australia. She was rather crowded on that voyage with no fewer than 64 aircraft and spare parts on board – 39 Sea Venoms, 22 Gannets for 816 and 817 Squadrons, two Bristol Sycamore helicopters and a single Avro 707A experimental delta wing jet aircraft! The journey to Sydney took nine weeks at an average speed of about 11 knots.

After first offloading the aircraft and equipment at Jervis Bay NSW, *Melbourne* arrived in Sydney to a tumultuous welcome on 9 May 1956. Five days later she was made flagship of the Australian fleet during a ceremony at the Garden Island Dockyard when Rear Admiral H M Burrell transferred his flag from HMAS *Sydney*. *Melbourne* remained flagship until she was paid off in 1982.

During a national broadcast on ABC Radio on 20 May, captain G C O Gatacre talked about *Melbourne's* working up period and her maiden voyage from Britain to Australia:

"After commissioning, *Melbourne* spent the next five months in English waters based on Plymouth or Portsmouth carrying out the large number of trials normally undertaken by a new ship, to test her machinery, equipment and performance. The trials in fact proved her as a ship and then proved her as an aerodrome.

"All ship's trials and flying trials were highly satisfactory, and although undertaken in the northern winter which brings a succession of gales, we completed a comprehensive programme without delay. When we needed reasonable weather for any particular trial, we were favoured by reasonable weather; when bad weather was of no consequence, a gale would come and go.

"At this stage our three air squadrons had formed ashore and were practising their airmanship and developing their skills with the new jet aircraft.

"During February [1956] we were honoured by a visit at Portsmouth by His Royal Highness the Duke of Edinburgh, who landed in his helicopter on our flight deck and spent three and a half hours on board. After inspecting the ship's company, he walked around the ship displaying a keen interest in everything and, of course, a sound professional knowledge.

"He then had lunch in the wardroom with a full sitting of officers. Before leaving the ship by helicopter, His Royal Highness expressed to me his congratulations on the appearance and cleanliness of the ship, and the smartness of her company.

"In March the ship embarked her air squadron personnel and at Glasgow loaded fully with aircraft and freight. Then we sailed for Australia.

"Our loading of aircraft on the flight deck and in the hangars filled all spaces in which

A Sea Venom from 805 Squadron shares Melbourne's *flight deck with a Fairey Gannet anti submarine aircraft. This pair made up the Fleet Air Arm's 'sharp end' until replaced by the A-4 Skyhawk and S-2 Tracker in 1967. (via Steve Long)*

Occupational hazards (top to bottom): WZ896 after overshooting at Nowra following engine failure in August 1957; WZ894 of 724 Squadron after another attempted dead stick landing in May 1959, this time at Jervis Bay; and WZ946 of 805 Squadron following brake failure aboard HMAS Melbourne *in March 1960. (via Steve Long)*

any numbers could have recreation so that it was necessary to plan in our voyage to make a brief stay at several ports so we could stretch our legs ashore. And thus it was we made short visits to Gibraltar, Naples, Malta, Aden and Colombo.

"After leaving Colombo the ship began to be cooled by the Trade Winds. We passed the Cocos Islands one afternoon and at the request of the Australian Administrator, steamed close alongside so the Australian community could see their new ship. Being beyond 'coo-ee' range, they greeted us by firing several coloured Verys lights.

"In Australia we were given a warm welcome in beautiful weather when we arrived in Fremantle, and although the less said about the weather in Melbourne the better pleased will be the Melbourne-ites; our welcome there was indeed a warm one, too.

"The occasion when a ship would berth on the port of the great city, whose name we proudly share, was of course one we had all been eagerly awaiting. And there at a civic reception at the town hall I received from the Lord Mayor a silver plaque – a generous gift from the citizens of Melbourne – to be mounted in HMAS *Melbourne*.

"Then on our way again, to Jervis Bay to unload aircraft for the Naval Air Station at Nowra and so finally here to Sydney ..."

Squadrons and Squabbles

Four Fleet Air Arm squadrons were the major operators of the Sea Venom during its decade of front line service with the RAN – Nos 808, 805, 816 and 724, the latter flying the aircraft from ashore at Nowra. Where appropriate, the early histories of these squadrons are summarised in the Sea Fury and Firefly sections of this book.

As noted above, 808 Squadron was the first to be equipped with the Sea Venom and was reformed in Britain in August 1955, disbanding at Nowra in December 1958 after undertaking seven periods of duty aboard HMAS *Melbourne* as the Fleet Air Arm's front line all weather fighter squadron. 808 Squadron Sea Venoms carried the identification letter 'Y' on their vertical tail surfaces.

808 Squadron's activities were taken over by 805 Squadron from late 1958, the famous former Sea Fury unit having been reformed at Nowra in August of that year and re-equipping with Sea Venoms which were identifiable by the letter 'M' on their tails. 808 Squadron served for 11 tours aboard *Melbourne* before being disbanded in June 1963, in the meantime surviving political attempts to effectively 'kill off' the Fleet Air Arm as a fixed wing operation. 805 Squadron was recommissioned in January 1968 equipped with A-4 Skyhawks.

In September 1959 the Australian government announced the Royal Australian Navy as a whole would concentrate on anti submarine warfare operations and that the defence forces generally would be standardised with those of the United States.

Sea Venom WZ903 photographed moments after leaving HMAS Melbourne's *catapult. This aircraft survives with Dennis Sanders at Chino, California, and is under restoration to fly again. (via Steve Long)*

Among the various proposals as they directly affected the Fleet Air Arm was one that it be disbanded in 1963 when the Sea Venoms and Gannets reached the end of their planned operational lives. It was considered that any likely replacement would not be capable of operating from *Melbourne*.

This attitude was modified slightly in 1960 when it was decided that *Melbourne* would continue in service but converted for anti submarine duties equipped with Westland Wessex helicopters. For the moment, the decision to end Fleet Air Arm fixed wing flying in 1963 remained in place.

The situation changed in that year when the threats associated with 'confrontation' with Indonesia were fully appreciated and the Sea Venoms and Gannets were given the go ahead to continue operations from *Melbourne* until 1967.

The Indonesian situation caused a general rethink of Australia's defences and of the role of the Fleet Air Arm, as did the Vietnam War. The result was an increase in spending and the placing of two important orders for fixed wing aircraft in the mid 1960s – 10 (later 20) Douglas A-4 Skyhawk fighter-bombers to replace the Sea Venoms and an initial 14 Grumman S-2 Trackers to replace the Gannets. Deliveries of these aircraft began in 1967 and they operated aboard *Melbourne* for 15 years before the politicians finally caught up with the Fleet Air Arm fixed wing operations!

The uncertainty of the FAA's situation in

Sea Venom WZ931 on the 724 Squadron flight line at Nowra in the 1960s. (via Steve Long)

(right) The Fleet Air Arm on parade at Nowra with 805 Squadron's Sea Venoms in the foreground and Gannets behind. (RAN)

1963 meant there was a gap between the disbanding of 805 Squadron in June of that year and re-introduction of the Sea Venom to carrier operations.

This occurred in July 1964 when four Sea Venoms were attached to the Gannet equipped 816 Squadron. Operating as B Flight of that squadron (identification letter 'M'), the Sea Venoms remained on strength until the squadron was disbanded in August 1967, preparatory to its reformation the following January equipped with Trackers.

The disbanding of 816 Squadron marked the end of the Sea Venom's front line and shipboard service with the Royal Australian Navy, although the type remained in use for a time after that.

724 Squadron

Shore support for the RAN Fleet Air Arm was based at HMAS *Albatross* at Nowra on the south coast of New South Wales. When not at sea, the operational squadrons operated from there.

One unit which was permanently based at Nowra was 724 Squadron, a training and miscellaneous squadron which listed among its numerous duties that of the Sea Venom Operational Conversion Unit and Headquarters Squadron.

724 Squadron had its origins in a Royal Navy squadron which was formed at Bankstown NSW in April 1945 equipped with Beechcraft 18 Expeditor and Avro Anson aircraft.

The squadron was disbanded in May 1946 but reformed as an RAN unit at Nowra in June 1955, its original aircraft complement comprising Wirraways, Sea Furies, Fireflies, Sea Venoms, Vampires and Gannets. Aircraft based at Nowra had the identification letters 'NW' on their tails. One of 724's main roles was to convert Sea Fury and Firefly pilots to the Sea Venom and Gannet, respectively.

The squadron's equipment was frequently added to and changed over the years, the addition of six Bristol Sycamore helicopters to the list in 1956 being the first major one. Shortly afterwards, the Sea Furies, Fireflies and Wirraways were transferred to 805 and 851 Squadrons, leaving the unit with Gannets and as an operational flying school with Vampires and Sea Venoms for jet conversion and all weather fighter courses.

In 1958, 724 Squadron became an all jet unit with Vampires and Sea Venoms fulfilling the roles mentioned above.

1959 saw the formation of a Sea Venom aerobatic team, the *Ramjets*. To some, this seemed an appropriate name as the trademark of the fuel supply company – Golden Fleece – was a ram. As discussed in the following 'Sea Venom Impressions' chap-

Line up of 805 Squadron Sea Venoms aboard HMAS Melbourne. *WZ930 in the foreground was one of several aircraft which soldiered on after most had been retired and was used for target towing and other fleet support duties. It was sold to the Tocumwal Historical Society in 1973. (via Steve Long)*

Aspects on landing a Sea Venom onto HMAS Melbourne. *The carrier had just five wires for the pilot to aim for, and as shown here, they weren't spread over an overly large distance! (RAN)*

Various Venoms photographed at Nowra in May 1963 (top to bottom): WZ907, WZ900 and WZ905. Of these, only WZ907 survives, in the Camden Museum of Aviation. WZ900 went over the side of HMAS Melbourne *in April 1966 and WZ905 was scrapped in 1967. (Richard Hourigan)*

ter, 724 Squadron had only four Venoms on strength at the time, making the running of a four-ship aerobatic team extremely difficult as normal training operations obviously took precedence and 100 per cent serviceability is not the norm for any fleet of aircraft, particularly military ones! A fifth aircraft was obtained for the squadron's use, helping the situation only slightly.

One of the *Ramjet's* tasks was to perform at the opening of the Avalon airfield in February 1961, while numerous other appearances helped fly the flag for the Fleet Air Arm.

The role of 724 Squadron expanded yet again in 1961 when 725 (anti submarine warfare training) Squadron was amalgamated with it, bringing the RAN's few remaining Sea Furies and Fireflies back into the fold along with Gannets, Austers and Dakotas. Officially, 724's role was now anti submarine and all weather operational training, communications, trials and fleet requirements.

The squadron took in other unit's aircraft as they disbanded over the next few years, including 805 Squadron's Sea Venoms, while the aircraft used by 816 Squadron's 'B' Flight aboard HMAS *Melbourne* from mid 1964 came from 724's stocks.

By then, the Sea Venom's operational life with the RAN was in its twilight and by the end of 1966 most of the surviving aircraft had been deleted. A few soldiered on with 724 Squadron as it again reverted to being an all jet conversion training squadron in that year in preparation for the arrival in 1967 of the first Douglas Skyhawks.

By mid 1967 only Sea Venoms WZ901, 938, 930, 935, 943 and 944 remained on strength, the last four of the above converted to the target towing role and instantly recognisable by their striped black and yellow undersurfaces. WZ935 and WZ943 were the last Sea Venoms deleted from the Royal Australian Navy's list, in June 1973.

Tours of Duty

The Sea Venoms' basic role during its years of service was to provide fighter protection for the ship and accompanying fleet, by day and by night. The aircraft was generally assessed as being quite successful as a night fighter, with a good (almost side-by-side) crew layout which encouraged close communication between the pilot and his observer/radar operator.

The AI Mk.17 radar was used by the observer not only for the tracking of targets but also for navigation. The Sea Venom was regarded as being a good instrument flying platform and it's interesting to note it was the only all weather fighter in Australian service until the Mirage joined the RAAF in 1964. Until then, the RAAF's front line fighter was the Sabre, an excellent aircraft but very much a day fighter nonetheless.

HMAS *Melbourne's* cruises usually involved exercises with other elements of the Australian armed forces and/or with those of friendly nations. Trips to New Zealand, the Far East and elsewhere were common, and from 1959 the ship and her aircraft

The old and the new. Sea Venom WZ895 poses with her replacement, the Douglas A-4G Skyhawk. These days, the Skyhawks have all gone but WZ895 survives with the RAN Historic Flight and will one day fly again. (via Steve Long)

made regular trips on exercises or as part of the Far East Strategic Reserve. These operations were based around Singapore and Malaysia but included voyages to Hong Kong and the Philippines.

Although the 1959 decision to end RAN fixed wing flying and make *Melbourne* a helicopter equipped anti submarine warfare (ASW) platform was eventually reversed (see above), the nature of the ship's operations did change in 1963 with the embarkation of 817 Squadron's Westland Wessex helicopters. The Wessexes replaced Gannets, although the other Gannet unit, 816 Squadron, continued with that aircraft for a while as well as the Sea Venom.

The exercises *Melbourne* and her aircraft participated in allowed the Sea Venom pilots to sometimes fly mock combat with other, usually superior aircraft, although not all the ship's patrols were routine.

Search and rescue work occasionally featured, one of the earliest examples occurring in August 1956 when the carrier joined in the search for survivors from the collier *Birchgrove Park* which had gone down north of Sydney.

Another regular excursion was in the mid 1960s when *Melbourne* escorted HMAS *Sydney* (by now used as a fast troop transport) to Vietnam. A less glorious chapter of *Melbourne's* career was recorded in February 1964 when the carrier collided with the destroyer HMAS *Voyager* off Nowra, badly damaging the bow section of the flagship and sinking the *Voyager* with the loss of 82 lives.

Melbourne's last activities with Sea Venoms aboard were recorded in 1967. In August of that year, 816 Squadron (Sea Venoms and Gannets) and 817 Squadron (Wessexes) were disembarked in preparation for *Melbourne's* departure for the USA the following month to pick up her new aircraft complement of Skyhawks and Trackers.

Early days in the restoration of Sea Venom WZ895 at the Naval Aviation Museum, with workers Pete Dikinson and Steve Long posing with their charge. (Steve Long)

Sea Venom Losses

Of the Royal Australian Navy's 39 Sea Venoms, 11 were written off whilst in service with the loss of nine aircrew. A summary of these losses is presented below:

WZ933, 8 August 1956: Crashed into sea after being catapulted off HMAS *Melbourne*, both crew killed. This was before the RAN's Sea Venoms were fitted with ejection seats. There was a known roll control problem with the Sea Venom immediately after take-off.

Sea Venom survivors (top to bottom): WZ937 in the care of the Naval Aviation Museum and photographed at Schofields NSW in 1981; WZ939 'up a pole' on the central coast of NSW; and the derelict WZ946 outside Chieftain Aviation at Bankstown airport NSW, photographed in 1993. (Stewart Wilson/Ross Gillett/John Moremon)

WZ896, 28 August 1957: Overshot runway at Nowra NSW after engine failure, crashed and damaged beyond economical repair.
WZ936, 17 September 1959: Crashed and burned on landing at Nowra NSW.
WZ894, 1 May 1959: Lost engine power and damaged beyond repair after dead stick landing at Jervis Bay airfield NSW.
WZ909, 20 May 1959: Crashed into sea off NSW coast while carrying out a photo exercise, crew lost. A CSIRO research vessel trawled up parts of the aircraft's wreckage (inner starboard wing, starboard boom/fin, tailplane, some fuselage components) in May 1988 from a depth of 3,655ft (1,114m).
WZ941, 11 February 1960: Crashed and burned on landing at Nowra NSW, crew killed.
WZ927, 15 June 1961: Crashed on landing at Nowra NSW, crew ejected.
WZ940, 5 July 1963: Crashed into Sydney Harbour after collision with WZ893 during flypast; pilot ejected, WZ893 returned safely to Nowra.
WZ911, 4 March 1964: Crashed Nowra NSW after the windscreen imploded.
WZ932, 3 December 1964: Crashed on landing at Nowra, pilot killed.
WZ900, 28 April 1966: Landing accident aboard *Melbourne* in Philippine Sea. Aircraft caught wire but the knuckle broke; insufficient power to go around so crew ejected; aircraft crashed into sea, observer killed.

Other accidents both on land and at sea caused damage to several Sea Venoms. Among the accidents aboard HMAS *Melbourne* over the years was one in November 1959 when the aircraft suffered an undercarriage collapse on hitting the round down; heavy landings caused airframe buckling in incidents in November 1959 and March 1960; also in March 1960 a Sea Venom suffered a nosewheel collapse and other undercarriage damage when it hit the safety barrier after failure of the brakes; and in a third incident in that same month, another aircraft was severely damaged when its starboard wing hit *Melbourne's* island. In all the above cases the aircraft were repaired and returned to service.

The Sea Venom suffered a couple of hiccups during its service with the RAN. The first was the discovery of a problem within the hydraulic system which operated the ailerons, the result being the potential for uncontrollable wing drop immediately after launching by catapult.

Very early in the aircraft's career, several aircraft were found to have developed cracks around the catapult hook area. Those in WZ937 were the first to be found, in March 1957 while the aircraft was serving aboard HMAS *Melbourne*. Repairs to the affected Sea Venoms took about a fortnight to perform (at Nowra), and in the meantime those aircraft were obviously grounded.

The RAN's Sea Venoms had very low utilisation; WZ937's records show that by the time it was 'paid off' in September 1963 it had flown only 1,180 hours, an average of just 156 hours per annum.

The end of the first part of WZ895's eventual restoration to flight with taxying trials at Nowra in 1983/84. The process of getting the Sea Venom into the air has proved to be a very lengthy one. (via Steve Long)

Disposal and Survivors

The majority of surviving Royal Australian Navy Sea Venoms were disposed of by the Australian Department of Supply in July 1966 and others followed after that. Of the 39 aircraft, 11 were written off in accidents and nine sold for scrap. Of the others, a healthy number were still in existence in 1993.

Three are undergoing restoration to airworthiness. WZ944 and WZ903 are with Dennis Sanders at Sanders Aircraft, Chino, California. WZ944 was originally sold to the Warbirds Aviation Museum, Mildura Vic and passed to Dennis Sanders in 1990. WZ903 has a similar history.

The third potential flyer is the Nowra Naval Aviation Museum's WZ895, for which the Australian civil registration VH-NVV has been reserved. This aircraft was retired from service with 816 Squadron in May 1966 after cracks were discovered in its lower port wing skin.

The aircraft was then relegated to instructional airframe duties at Nowra before being transferred to the Fleet Air Arm Museum in 1974 for static display. Restoration work to ground running status began in 1981, the first ground run being carried out in September 1982. Taxying trials began in February 1983 and the long slow process which will hopefully result in this Sea Venom flying with the RAN Historic Flight began in August 1985.

Other surviving RAN Sea Venoms in 1993 include the fuselage of WZ897 at the Camden Museum of Aviation in New South Wales and the complete WZ907 at the same place; WZ898 and WZ910 at the Queensland Air Museum, Caloundra; WZ901 at the Moorabbin Air Museum, Victoria; WZ904 is with Sid Beck at Townsville, Queensland; WZ930 resides with the Tocumwal Historical Society in NSW; WZ931, 935, 937 (the static restoration of which was completed in 1993) and 943 are at the Naval Aviation Museum, Nowra, the latter after spending a dozen years up a pole in a Nowra park; WZ939 is in storage with the Newcastle Military Museum, NSW; and the unfortunate WZ946 has spent nearly a quarter of a century outside Chieftain Aviation at Sydney's Bankstown Airport suffering the ravages of the elements, vandals and other hazards.

WZ931 at the Naval Aviation Museum in company with a Westland Scout, Bristol Sycamore and Fairey Gannet. (RAN)

FLYING THE DE HAVILLAND SEA VENOM

Commodore Norman Lee gives his impressions of the Sea Venom 'from the cockpit'.

When it came time for the RAN to replace the Sea Fury, a number of aircraft were considered but the obvious choice, considering the size of HMAS *Melbourne*, was the De Havilland Sea Venom, a two seat all weather jet fighter.

The aircraft entered service at the same time as the Gannet, the pair forming the second generation of fixed wing aircraft in the RAN. The Sea Venom was powered by a De Havilland Ghost centrifugal gas turbine and its twin boom configuration made it obvious that it was developed from the Vampire.

The aircraft was not initially fitted with ejection seats, which made the cockpit a little cramped when it was eventually installed as a mod. The pilot and observer sat side-by-side with the latter staggered slightly aft. The ailerons were hydraulically powered with the stick to aileron throw adjustable through a gearing control in the cockpit. This gave a range of half aileron for full stick to full aileron for full stick.

Four 20mm cannon were mounted in the nose and rockets could be carried under the wings. Surprisingly, the aircraft was not capable of carrying bombs.

The view from the cockpit was adequate, but the canopy coaming intruded particularly to starboard. The large flat windscreen was not the best in heavy rain, but there was a removable clear view side panel if things became too fraught. Trim control was in time honoured tradition by trim wheels in a trim box below the pilot's left hand.

Cockpit layout was the usual compromise but one soon got used to it, however care needed to be exercised with the HP cock and speed brake lever if one was flying Vampires concurrently as they were transposed in the two aircraft. Starting was by cartridge and straight forward.

The Sea Venom sat very solidly on the ground and was the least crosswind affected aircraft I have ever flown. Takeoff and climb were straight forward with the aircraft accelerating very rapidly to climb speed.

In my first battle drill climb in the aircraft I found that none of us could keep up with the squadron CO. He was constantly badgering us to keep up but we found it impossible. The same thing happened on the next occasion and I became suspicious that perhaps we weren't at fault as I had discovered that he was always at pains to fly the same aircraft. My primary task in the squadron was as the QFI but I also happened to be the Senior Pilot, responsible for the overall squadron programming. I tasked myself for a solo sortie in the suspect aircraft and timed it in the climb, discovering, not to my surprise that its performance far exceeded a standard Venom.

Now, the method of trimming the Ghost for power was by varying the tailpipe diameter. When the engine came off test it

Climbing into the Sea Venom, in this case the Naval Aviation Museum's WZ895 preparatory to taxying trials in 1983. (via Steve Long)

THE SEA VENOM COCKPIT

(left) A general view of the Sea Venom cockpit, with the observer's radar viewer on the far right. (Bruce Chapman)

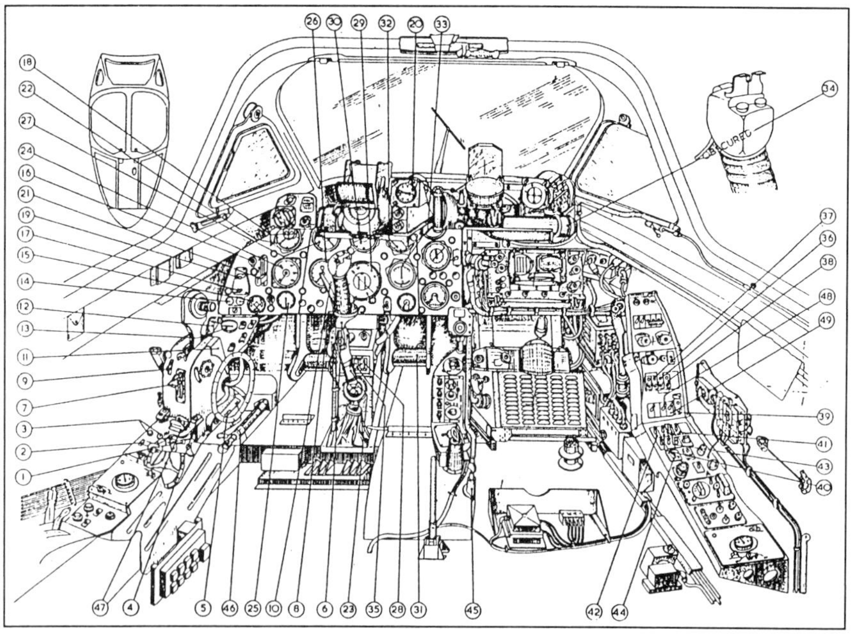

FIG.1 FLYING CONTROLS & EQUIPMENT: 1 - arrestor hook lever; 2 - RATO gear jettison switch; 3 - wing lock indicator; 4 - hook 'down' warning lamp; 5 - elevator trim control wheel; 6 - control column; 7 - flap selector lever; 8 - brake lever parling catch; 9 - elevator trim tab indicator; 10 - brake lever; 11 - RATO firing button; 12 - dive brake lever; 13 - hydraulic brake triple gauge; 14 - flap indicator; 15 - undercarriage position indicator; 16 - RATO warning lamp; 17 - RATO safety switch; 18 - E2A compass; 19 - RATO master switch; 20 - accelerometer; 21 - undercarriage selection lever; 22 - Machmeter; 23 - flight instrument switch; 24 - radio altimeter; 25 - cabin altimeter; 26 - altimeter; 27 - air speed indicator; 28 - press. head heater switch; 29 - Mk.G4F gyro compass; 30 - artificial horizon; 31 - rudder pedals; 32 - rate of climb indicator; 33 - turn and slip indicator; 34 - outside air temperature; 35 - battery isolation switch; 36 - standby inverter circuit breaker; 37 - external lights circuit breaker; 38 - main inverter circuit breaker; 39 - port aileron tab circuit breaker; 40 - navigation lamps switch; 41 - identification lamp selector switch; 42 - navigation lights dimmer switch; 43 - undercarriage shore/carrier switch; 44 - landing light selector switch; 45 - powered aileron selector valve; 46 - aileron spring strut; 47 - aileron trim tab switch; 48 - main inverter ground test switch; 49 - main inverter reset switch.

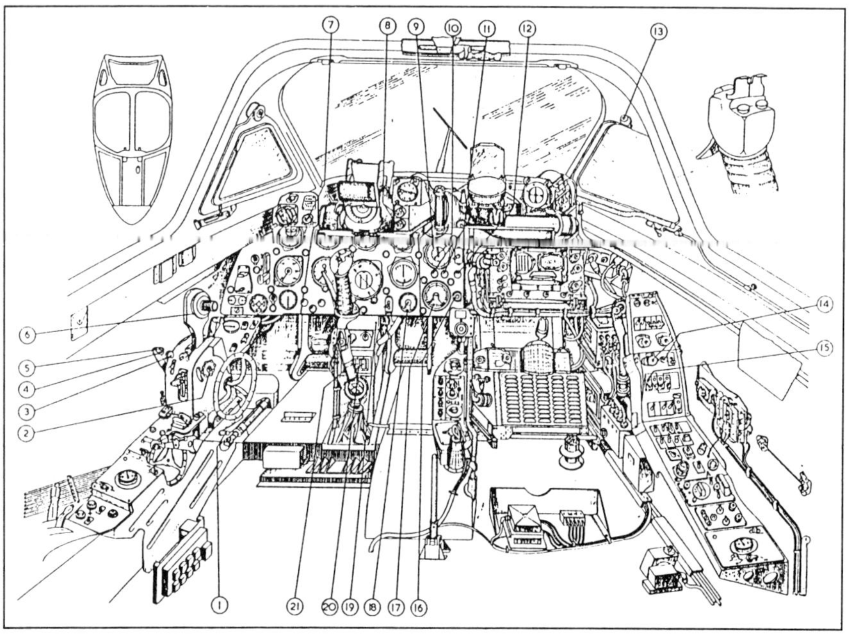

FIG.2 ENGINE CONTROLS AND EQUIPMENT: 1 - LP fuel cock lever; 2 - throttle friction damper; 3 - HP fuel cock lever catch; 4 - flight relight button; 5 - HP fuel cock lever; 6 - throttle lever; 7 - hydraulic flow failure warning lamp (pilot); 8 - generator failure warning lamps (pilot); 9 - tachometer; 10 - JPT gauge; 11 - fuel transfer indicators (tip tanks); 12 - fuel pressure warning indicator; 13 - hydraulic flow failure warning lamp (navigator); 14 - generator failure warning lamps (navigator); 15 - fuel pump circuit breaker; 16 - fuselage tank contents push switch; 17 - fuel contents gauge; 18 - oil ternperature gauge; 19 - engine starter push switch; 20 - fuel booster pump switch; 21 - engine starter master switch.

FIG.3 OPERATIONAL CONTROLS AND EQUIPMENT: 1 - pilot's oxygen regulator panel; 2 - tip tank fuel jettison switch; 3 - press to transmit button (pilot); 4 - GGS range control twist grip; 5 - Mk.5 gyro gunsight; 6 - collimator; 7 - AI Mk.17 unit Type 900; 8 - UHF homing indicator; 9 - RP/bomb switch safety flap; 10 - gun firing trigger safety flap; 11 - RP/bomb firing switch; 12 - camera switch; 13 - gun firing trigger; 14 - oxygen contents gauge; 15 - AI Mk.17 supply panel; 16 - VHF control unit; 17 - IFF aerial selector switch; 18 - gun camera sunny/cloud switch; 19 - gun camera master switch; 20 - GGS circuit breaker; 21 - IFF control unit; 22 - navigator's oxygen regulator panel; 23 - G4F compass corrector control panel; 24 - press to transmit switch (navigator); 25 - press to mute switch (navigator); 26 - AI Mk.17 control unit Type 901; 27 - AI Mk.17 indicator CRT100; 28 - RP pairs/salvo switch; 29 - armament master switch; 30 - UHF control panel; 31 - VHF/UHF audio control switch; 32 - GGS control selector/dimmer; 33 - cabin pressure warning lamp; 34 - radar dousing switch.

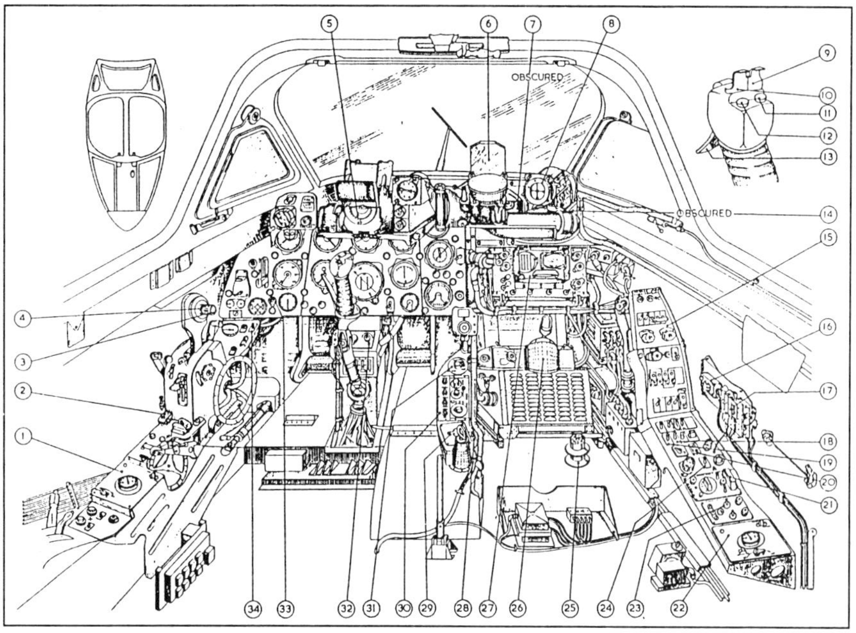

FIG.4 MISCELLANEOUS CONTROLS AND EQUIPMENT: 1 - red lamps master switches; 2 - wing fold selector lever; 3 - wing fold lock lever; 4 - pilot's panel lamp dimmer switches; 5 - windscreen wiper control knob; 6 - temperature control handwheel; 7 - hatch gun locking pin stowage; 8 and 9 - card holders; 10 - hatch external release handle; 11 - hatch jettison gun lock handle; 12 - pilot's direct vision window; 13 - hatch internal opening handle; 14 - windscreen wiper; 15 - windscreen deicing pump control; 16 - punkah louvre; 17 - hatch balance arm release catch; 18 - fuel pump test socket; 19 - filler cap spanner; 20 - inspection lamp socket; 21 - navigator's panel lamps dimmer switch; 22 - fuel pump test switch.

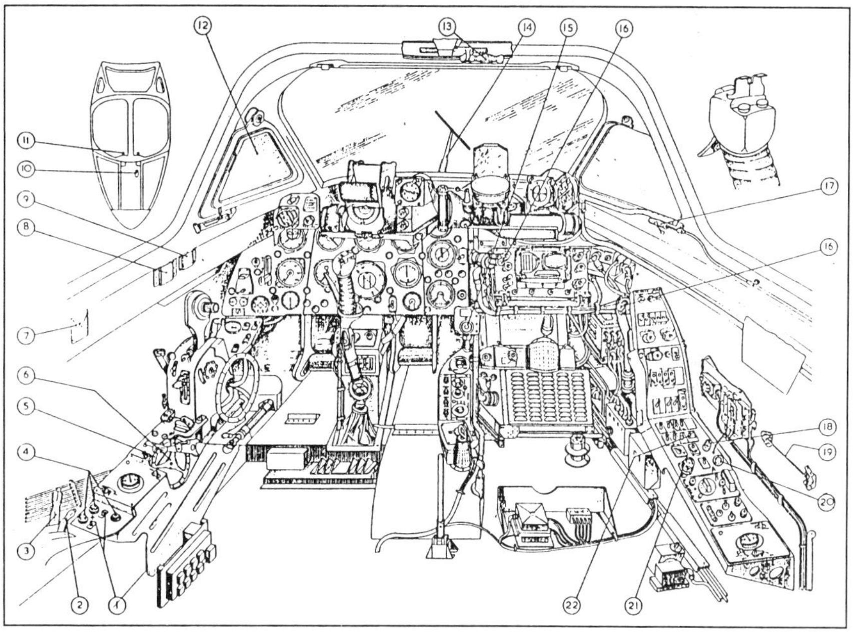

FIG.5 EMERGENCY CONTROLS AND EQUIPMENT: 1 - port aileron trim cutout switch; 2 - port aileron tab selector switch; 3 - tip tanks fuel jettison switch; 4 - port aileron neutral warning lamp; 5 - cabin pressure warning lamp; 6 - fire warning lamp and extinguisher push switch; 7 - hydraulic flow failure warning lamp; 8 - generator failure warning lamps (pilot); 9 - collimator platform release; 10 - emergency lamp switch; 11 - hydraulic flow failure warning lamp (starboard); 12 - generator failure warning lamps (navigator); 13 - canopy hatch jettison handle; 14 - emergency hydraulic hand pump; 15 - turn and slip indicator emergency battery switch; 16 - main inverter failure indicator; 17 - aileron power failure warning indicator; 18 - undercarriage emergency retraction switch.

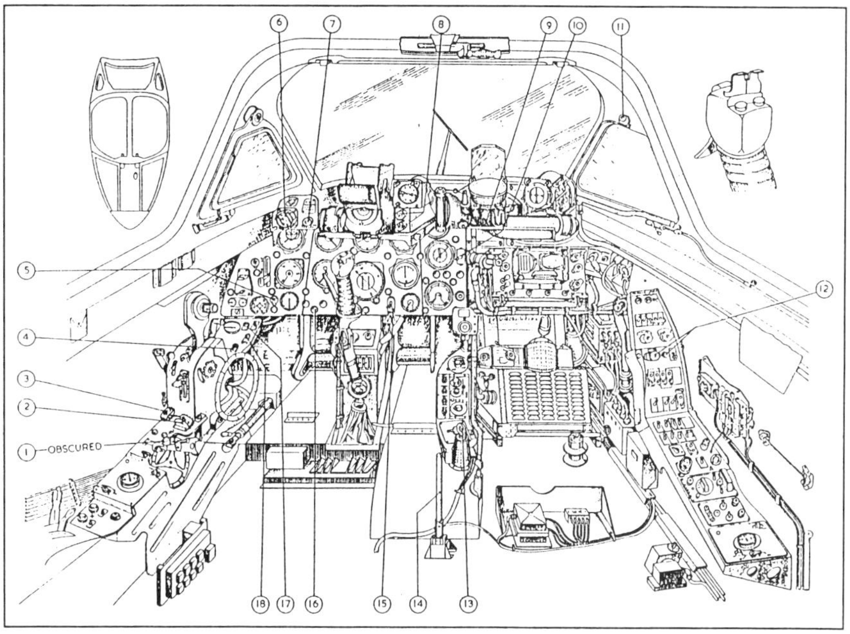

Flaring right over the piano keys (navy pilots obviously don't like to 'waste' runway!) is this RAN Sea Venom fighter of the fifties. (bottom) Probably the ultimate piston engine fighter, the gutsy Hawker Sea Fury served the RN and RAN well for many years. Grizzled naval aviation veterans still go watery-eyed at the thought of having flown these exceptional aeroplanes during the early days of their careers. (Eric Allen)

Sea Venom WZ895 (top) has been restored by the Naval Aviation Museum and is gradually working its way towards airworthiness. Year earlier, Sea Venom WZ897 is seen at work aboard HMAS Melbourne. This aircraft survives in the Camden Museum of Aviation. (Steve Long/Ross Gillet)

(top) Black smoke pours from the cartridge start system of this Sea Venom. The Sea Venom replaced the Sea Fury in the fighter role and in turn was ultimately replaced by the Douglas A-4G Skyhawk. Like its RAAF cousins, the FAA used the Vampire as its jet trainer till the delivery of Macchi jets in 1970. (Eric Allen)

A head on view of the Queensland Air Museum's WZ898 clearly shows the asymmetric, bulged canopy of the Sea Venom. (Ron Cuskelly via Steve Long)

would be noted as requiring a certain size tailpipe. A quick check of the aircraft log book soon showed that it had been fitted with the wrong size pipe and hence was producing above specification thrust. The Ghost for some reason ran well below maximum permissable JPT hence it is understandable that no one had queried it before. I snagged the aircraft and we had no more troubles, but I must admit that I wasn't popular in certain quarters!

I then had a break of two years with the RN and on return to command the All Weather Fighter training squadron, was informed that our task for the next three months was to form and work up an aerobatic team for a display at an international air convention at Avalon.

This was at the time of yet another rundown of the FAA and the squadron only had four Venoms on strength. I managed to win a fifth but it is not easy to run a four plane team with only five aircraft!

I had not flown formation aerobatics before, except for the odd play in Fireflies, but three of my pilots had formed part of an earlier team. Under the circumstances, I decided the safest thing for all of us would be if I led the team myself. The RN squadron with which I had been serving had flown a four plane team of Seahawks, so I was familiar with the various routines that had been developed overseas.

I discussed with my team members the speeds previously flown in looping and rolling manoeuvres, and then flew a pro-

With Macchis and Skyhawks in the background, Sea Venom WZ895 starts its Ghost engine with the characteristic pall of black smoke from the cartridge starter. (via Steve Long)

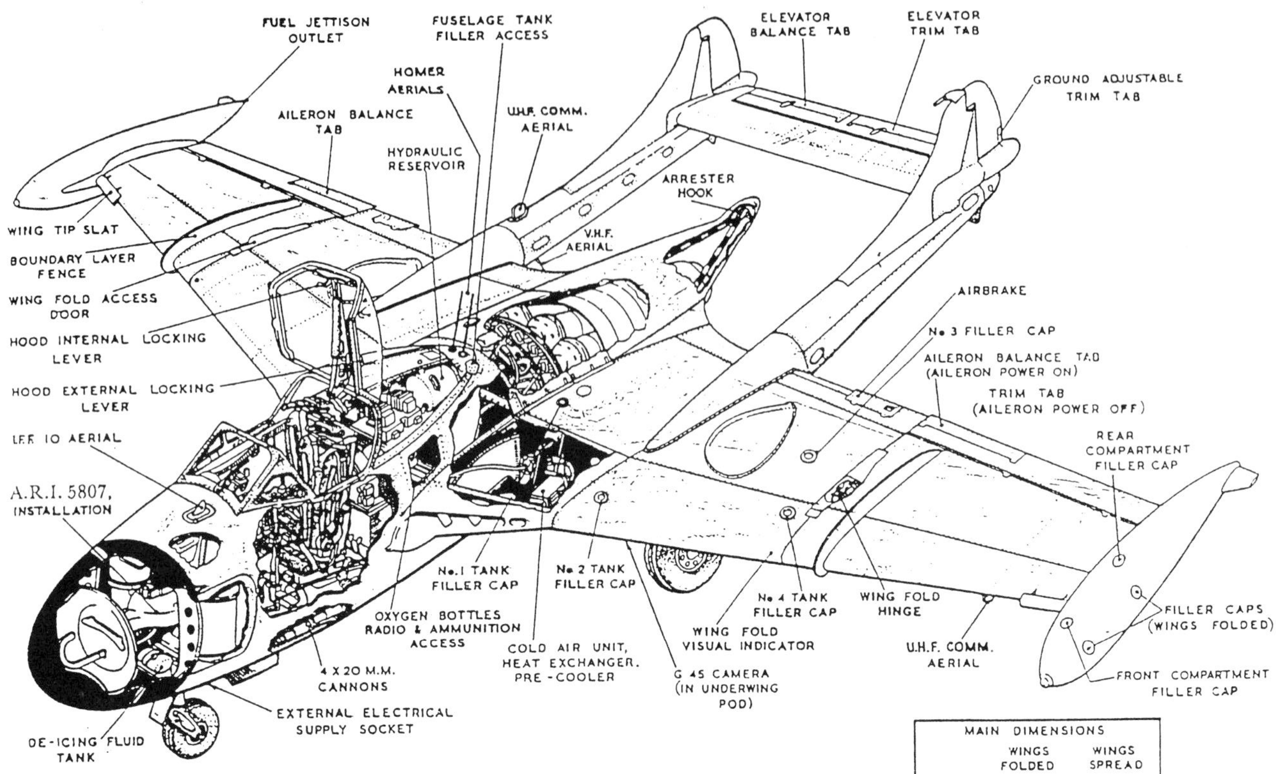

SEA VENOM FAW.53
(WITH EJECTION SEATS)

MAIN DIMENSIONS	WINGS FOLDED	WINGS SPREAD
SPAN	23' 0"	42' 10"
HEIGHT	9' 7" (13' 11" DURING SPREADING)	8' 6"
LENGTH	36'-8"	

posed routine solo. It was immediately apparent that they had been flying too fast, taking up too much airspace. After a few sorties we managed to knock 50 knots off all speeds and it was generally agreed that it was much more comfortable, with less loading in the vertical plane. The final routine evolved consisted of a four plane takeoff, followed by a double loop, two barrel rolls, a half loop and bomb burst, followed by a thread-the-needle and a reform and four plane formation landing. It was not a long routine, but then we hadn't been allocated very much display time.

Because of lack of aircraft it was rare that we managed to practise with all four team members together, which meant that I had to do a lot more flying than the others. This was no problem as I am a self confessed hour hog and really enjoyed the challenge of working the team up under somewhat trying circumstances.

The formation takeoff was made in box configuration with a positive liftoff and the boxman calling airborne. We then climbed through a full circle to a height from which we could do a double loop along the full length of the runway. Here we struck an unexpected problem; my loops weren't quite vertical and we were stepping aside several runway widths by the time we had finished the second loop. The cause was obvious, the Sea Venom had the stick cranked to the right so that you could see the instrument panel and I was instinctively pulling it straight back, feeding in a small amount of aileron. The problem was soon sorted out and we managed to do straight loops from thereon in.

The next manoeuvre was a barrel roll off the deck, followed by another returning back across the airfield. It was several years later that the RAAF suffered the loss of a Vampire aerobatic team at East Sale in, I believe, a barrel roll. I always made certain that I pulled up at least 30 degrees above the horizontal before initiating the roll, and that we were rolling with positive rudder. Aileron alone could leave the wingmen "hung up" at the top with the roll tending to fade out.

I will be corrected I know, but I believe I can lay claim to introducing the vertically downwards bomb burst to the Australian scene. We would climb to looping height following the second barrel roll and enter a loop along the axis of the main runway. I would call the burst just before we were pointing straight down. I would continue on straight ahead completing the loop, with the wingmen turning 90 degrees and the boxman 180 degrees, and then completing the loop. I would call the second loop and half roll to initiate the thread-the-needle. This is always a crowd pleaser as it looks dangerous whereas it is perfectly safe providing you stick to the rules. Our rules were that the leader and boxman returned down the runway keeping to the right, and on the deck. The two wingmen kept to the right of the tower and flew at mid tower height. We occasionally achieved the perfect crossing but there were too many factors involved to guarantee it every time.

Should we be landing on completion of the display, we would form up and do a box landing with the boxman calling the cut.

Towards the end of our work up, the wingmen were tending to get a little close to the extent that I could feel aerodynamic interference on my aircraft. The Commander (Air) at the time became concerned at the tightness of the formation and indicated that I should direct them to move out. On that very day, upside down over the married quarters, half way through a barrel roll, I felt an almighty thump through the rudders. My immediate thought was that I had snapped a rudder cable. My Observer said a rude word and the formation became a little ragged as we rolled out.

A check of all the aircraft could find nothing wrong , but I called it off nevertheless. Back in dispersal it became immediately obvious what had happened from the blue and gold tip-tank marking that had been transferred to the bullet fairing around my fin and rudder. My No 3 wingman had got too close and his tip tank had grazed my bullet fairing, sucking my rudder across due to the venturi effect. There was absolutely no damage to the bullet fairing and it was only necessary to wipe the paint with thinners to remove all trace of the incident. I then agonised over whether I should in-

Pilot's eye view of 'The Ramjets' aerobatic team in action. (via Steve Long)

form Commander (Air), finally deciding to do so the following day as we flew down to Avalon together in a Vampire; there was not much he could do at 30,000 feet!

Our great day came and with it turbulence and a low cloud base. We flew our number two sequence which deleted looping manoeuvres but it was a terrible let down after all our practising.

This was the early 60s and outside broadcasts on TV had not yet been developed. Film would be rushed to the TV station to be broadcast an hour or so later. As a result we had time to land at Laverton, our base for the display, and then see ourselves performing on the mess TV!

Shortly after our return to Nowra, we were scheduled to put on a display for a visiting parliamentary committee. Once again the weather upset our plans to the extent that I could only safely fly the boxman for a few rolling manoeuvres. We finished off with a pairs run close by the tower with everything down. The committee was standing on the tower balcony so in time honoured naval tradition I gave them a salute. That evening in the mess we met the committee and one of their number in conversation admitted that he had watched our performance very keenly because he had not wanted to miss the prang should it occur.

In due course I left the squadron to do some "fishheading" (general service sea time) and the team reformed with a new boxman and one of my old wingmen leading. If you're old enough you may recall a Sea Venom spearing into Sydney Harbour with the pilot ejecting in the nick of time. This happened when the team was putting on a display over the Harbour and the newbox man clipped one of the wingmen as they split in the bomb burst. The wingman got back to NAS Nowra minus a large section of his tail and that was the end of aerobatics over the city!

Never having qualified in the all weather role I thought I had better do so as the CO of the all weather training squadron! As this entailed mostly night flying it didn't interfere with the work up of the aerobatic team. It was a fairly condensed course as I was the only pupil and we used a Vampire as the target aircraft to save the Venoms. Having just spent two years in day fighters I found the change a little difficult to get used to, one had to trust the observer, particularly close in.

In summary, the Sea Venom was a good solid aircraft with no vices. It was pleasant and easy to fly and had a relatively good safety record. I can only recall being embarrassed once in a Sea Venom and that was on a land away to Amberley when I forgot to close the jettison valves in the tip tanks and on refuelling, the aircraft poured fuel all over the tarmac. I had obviously learnt nothing from my Oxford experience thirteen years previously.

Sea Venom WZ931, preserved at the Naval Aviation Museum, Nowra. (via Steve Long.

RAN DE HAVILLAND SEA VENOMS

Abbreviations: w/o – written off; DoS – Department of Supply; deliv – delivery; dbr – damaged beyond repair; C/No – constructor's number; NAM – Naval Aviation Museum; TT – target tug.

Notes: All aircraft were collected from the United Kingdom and delivered to Australia aboard HMAS *Melbourne.*

Serial prefixes applied to Royal Australian Navy aircraft 1964, of which the Sea Venom was allocated N4 followed by a number reflecting the original British serial (eg WZ987 became N4-987). These are noted where appropriate in the 'Remarks' column.

Note the Side Numbers listing is not necessarily complete.

All RAN Sea Venoms were FAW Mk.53s and all were handed over to the service in Britain on 27 February 1956.

Serial	C/No	Side Nos	Deliv	Disposal/Remarks
WZ893	12750	882	02/56	sold by DoS 07/66, scrapped Bankstown NSW 05/67
WZ894	12751	205/806	02/56	dbr dead stick landing Jervis Bay NSW 05/59
WZ895	12752	804/870	02/56	extant Naval Aviation Museum, Nowra NSW, under rebuild to fly as VH-NVV
WZ896	12753	206	02/56	crashed Nowra NSW 08/57, engine failure
WZ897	12754	801/864	02/56	(N4-897) fuselage extant Camden Aviation Museum NSW
WZ898	12755	874	02/56	to Queensland Air Museum 1979
WZ899	12756	801/885	02/56	sold by DoS 07/66, scrapped Bankstown NSW 05/67
WZ900	12757	866	02/56	landing accident HMAS *Melbourne* Philippine Sea 04/66, over side
WZ901	12758	808/810 967/875	02/56	(N4-901) to Moorabbin Air Museum Vic 1971
WZ902	12759	807/884	02/56	sold by DoS 07/66, scrapped Bankstown NSW 05/67
WZ903	12760	812/871	02/56	sold by DoS 12/70, extant Sanders Aircraft, Chino, California USA
WZ904	12761	804/868	02/56	(N4-904) sold Sid Beck, Townsville Qld 10/74
WZ905	12762	863/886	02/56	sold by DoS 07/66, scrapped Bankstown NSW 05/67
WZ906	12763	804/880	02/56	sold for scrap by DoS 07/66
WZ907	12764	808/886 875/864	02/56	sold 07/66, to Camden Aviation Museum NSW
WZ908	12765	864/887	02/56	sold by DoS 07/66, scrapped Bankstown NSW 05/67
WZ909	12766	987/865	02/56	crashed into sea off NSW 05/59
WZ910	12767	808/867	02/56	(N4-910) sold by DoS 12/70, to Queensland Air Museum
WZ911	12768	810/887	02/56	crashed Nowra NSW 03/64 and w/o
WZ927	12769	867	02/56	crashed on landing Nowra NS 06/61
WZ928	12770	866	02/56	sold by DoS 07/66, scrapped Bankstown NSW 05/67
WZ929	12771	800/810 803/881	02/56	sold by DoS for scrap 07/66, extant Parafield SA
WZ930	12772	806/865	02/56	(N4-930) cvtd TT, sold Tocumwal Historical Society 12/73
WZ931	12773	887/877	02/56	to NAM, Nowra NSW
WZ932	12774	864	02/56	crashed on landing Nowra NSW 12/64
WZ933	12775		02/56	crashed into sea, catapult accident HMAS *Melbourne* 08/56
WZ934	12776	208/864	02/56	sold by DoS for scrap 09/67
WZ935	12777	210/878	02/56	(N4-935) cvtd TT, sold 12/73, extant NAM, Nowra NSW
WZ936	12778		02/56	crashed on landing Nowra NSW 09/57
WZ937	12779	211/874 802/862	02/56	Naval Aviation Museum, Nowra NSW
WZ938	12780	208/872	02/56	to firefighting training Nowra 07/68
WZ939	12781	803/879	02/56	(N4-939) to Tuggerah Lakes War Museum 1971
WZ940	12782	809	02/56	crashed into Sydney Harbour after collision with WZ893 07/63
WZ941	12783	866/975	02/56	crashed on landing Nowra NSW 02/60
WZ942	12784	809	02/56	to firefighting training Nowra 01/63
WZ943	12785	805/865	02/56	cvtd TT, donated to Nowra city 12/72, to Naval Aviation Museum 1984
WZ944	12786	873	02/56	cvtd TT, extant Sanders Aircraft, Chino California USA, rebuild to fly
WZ945	12787	805/866	02/56	sold by DoS 07/66, scrapped Bankstown NSW 05/67
WZ946	12788	807/817 809/869	02/56	to Chieftain Aviation, Bankstown NSW 09/69

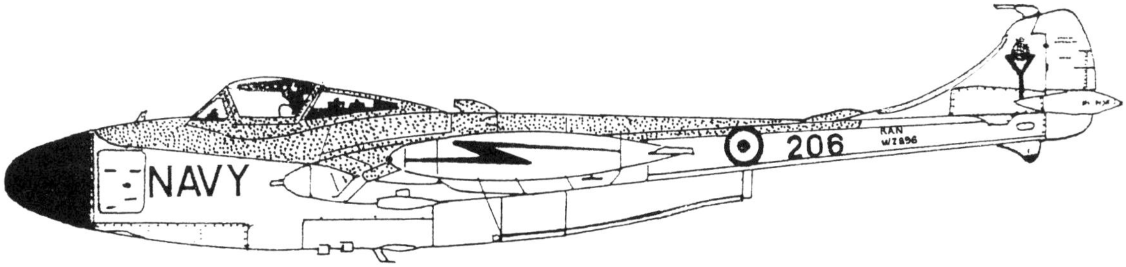

Sea Venom WZ896/206Y aboard HMAS *Melbourne*, late 1950s
Standard scheme of extra dark sea grey upper surface with sky undersurface. Black lettering, numbers, serial and code. Squadron badge on fin. Early type roundels in the standard six positions.

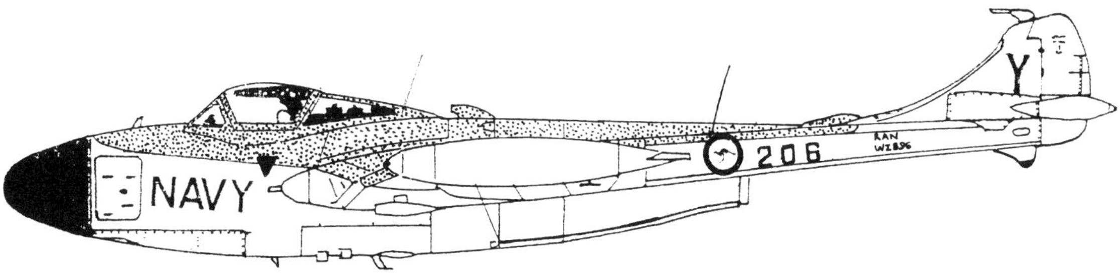

Sea Venom WZ896/206Y aboard HMAS *Melbourne*
Standard scheme. Kangaroo roundels in six positions. Insignia red ejection warning triangle below cockpit.

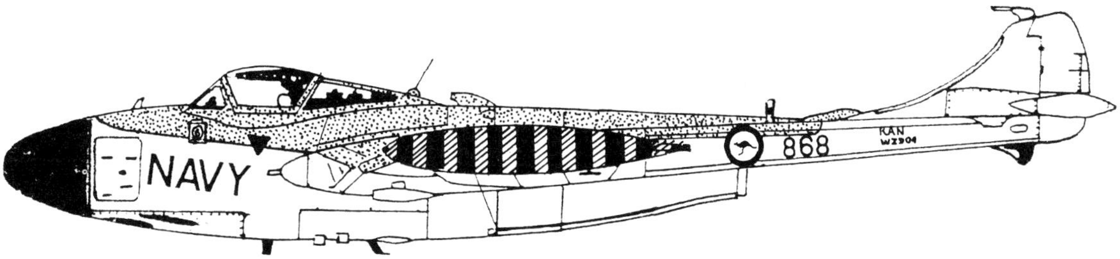

Sea Venom WZ904/868 used for training
Standard scheme. Black and yellow fuel tanks. Squadron badge below windscreen. Black tail bumper with yellow leading edge.

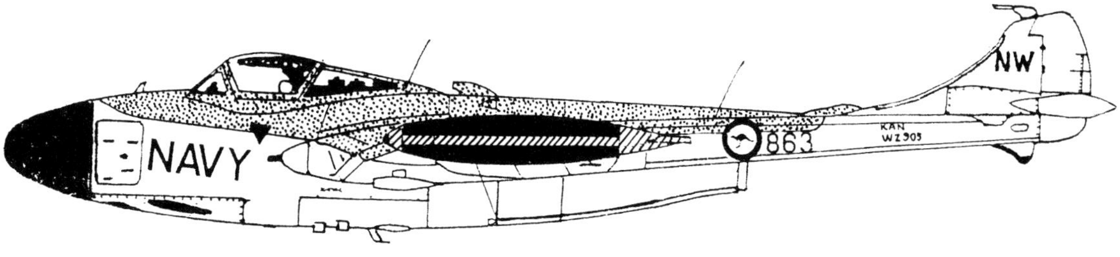

Sea Venom WZ905/863NW of VS 816
Standard scheme. Dark green wing tip tanks with yellow arrow. Black tip on tail bumper with black leading edge.

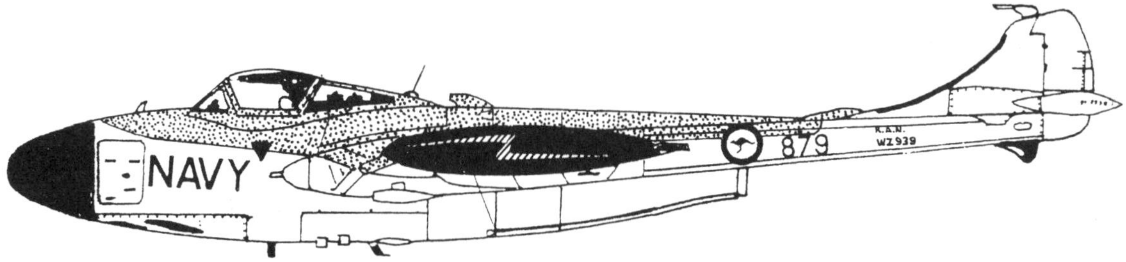

Sea Venom WZ939/879 of VC 724
Standard scheme. Black leading edge on fin. Insignia blue wing tip tanks with yellow flash. Insignia blue tail bumper with yellow leading edge.

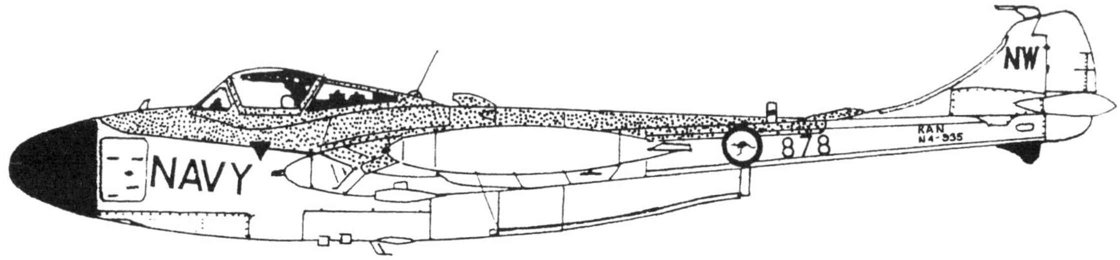

Sea Venom N4-935/878NW
Standard scheme. Black tail bumper.

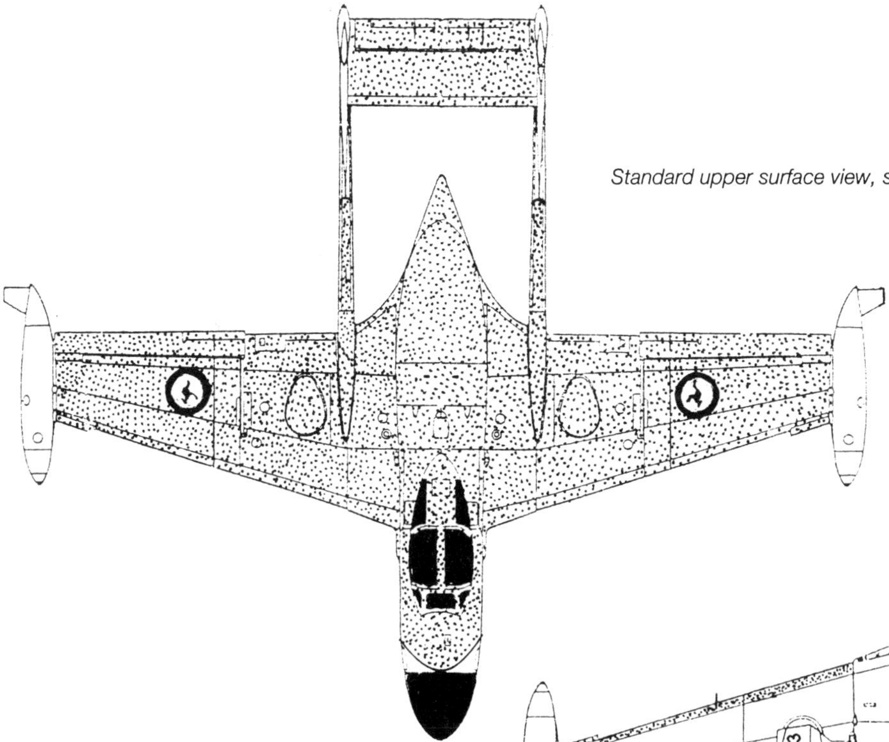

Standard upper surface view, showing location of roundels.

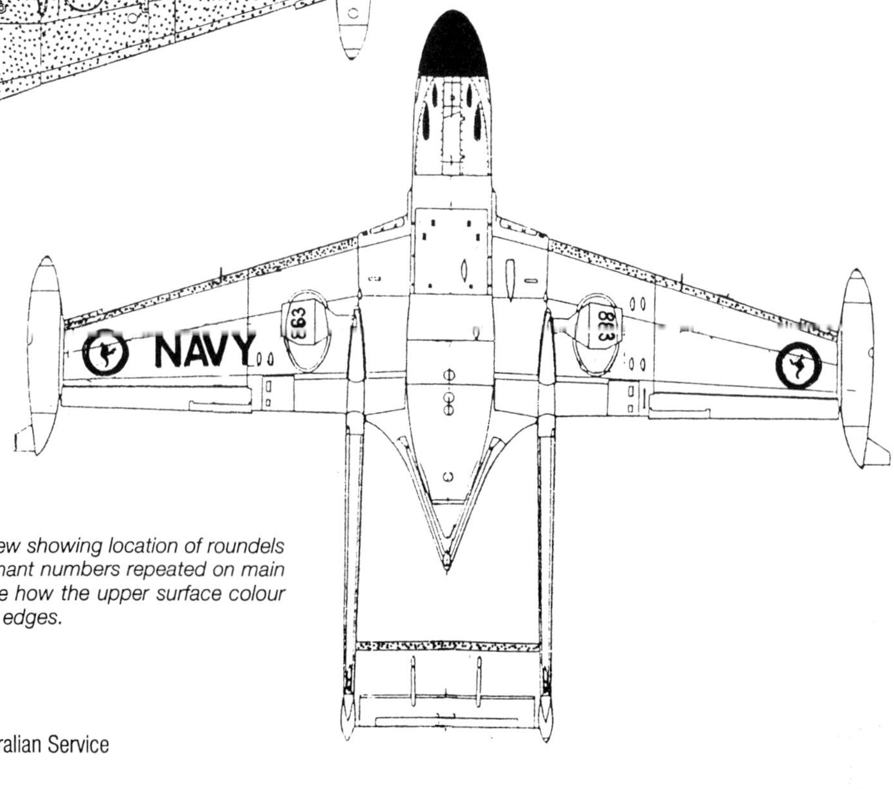

Standard undersurface view showing location of roundels and black NAVY and pennant numbers repeated on main undercarriage doors. Note how the upper surface colour wraps around the leading edges.

CELEBRATING THE AIRCRAFT THAT MADE AUSTRALIAN MILITARY AVIATION HISTORY

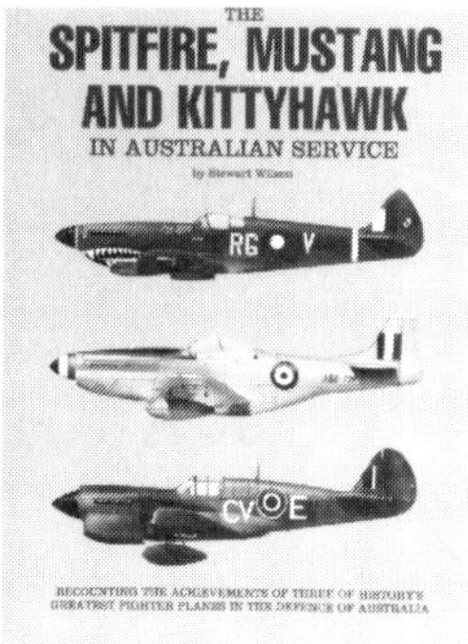

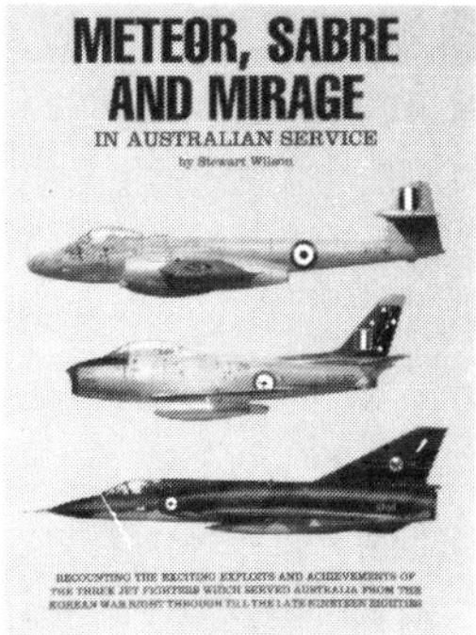

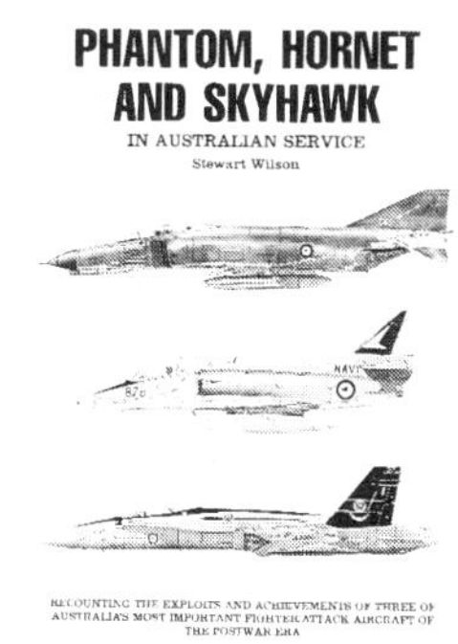

EACH BOOK IS ESSENTIALLY THREE BOOKS IN ONE!

★ Individually narrates the exciting development history of each particular aircraft including their entry to general service worldwide.
★ Documents in detail each aircraft's adaptation to Australian service and describes their war record, where applicable, in full.
★ Unique tables document each aircraft individually, listing its A series registration, individual version or batch number, where applicable, date of delivery and, most importantly, its eventual fate. This never before available data will provide a unique source of reference material for years to come.
★ Over 100,000 words of text backed up by hundreds of photos in each book, most of which have never before been published. These excellent books are worth the modest purchase price for the pictorial content alone.

POST TO: AEROSPACE PUBLICATIONS P/L, PO BOX 3105, WESTON CREEK 2611. TEL: (06) 288 1677, FAX: (06) 288 2021

Soft cover $19.95 Hard cover $29.95, plus $4.95 p&p for first book and $2.00 for each additional book.

($A30 or $A40 respectively for overseas seamail orders, p&p included).

★ ★ ★ Save with soft cover multiple orders (domestic sales only) ★ ★ ★
Any 3 titles $60, any 4 titles $75, any 5 titles $90, includes p&p

Name .... Address ....

.... P/C .... ☎ ....

Soft	Hard		
❒	❒	Spitfire, Mustang & Kittyhawk	$ ....
❒	❒	Meteor, Sabre & Mirage	$ ....
❒	N/A	Lincoln, Canberra & F-111	$ ....
❒	N/A	Beaufort, Beaufighter & Mosquito	$ ....
❒	N/A	Dakota, Hercules & Caribou	$ ....
❒	❒	Catalina, Neptune & Orion	$ ....
❒	❒	Wirraway, Boomerang & CA-15	$ ....
❒	❒	Boston, Mitchell & Liberator	$ ....
❒*	❒	Anson, Hudson & Sunderland	$ ....
❒*	❒	Phantom, Hornet & Skyhawk **	$ ....
Plus post & pack (see above)			$ ....
		TOTAL	**$ ....**

❒ I enclose my cheque/MO for $ .... payable to Aerospace Publications P/L **OR** please charge to:

❒ Bankcard ❒ MasterCard ❒ AmEx ❒ Visa

Card No: ....

Expiry Date: .... Signature ....

Please allow 3 weeks for delivery

* These soft cover editions also available at most major newsagencies but all other editions ONLY available by direct subscription.

** Available late April 1993.

WINGS ACROSS THE SEA

by Ross Gillett

This 164 page book features the complete history of over 70 years of Australian naval aviation, through both peace and war. An excellent and informative narrative in addition to being of immense reference value. It has complete tabulated data on every aircraft to have served with the RAN including individual squadron histories and a chronology of major RAN FAA events.

Wings Across The Sea is illustrated with more than 290 photos including 34 in glorious colour, the vast majority of which have never before been published.

TO: AEROSPACE PUBLICATIONS P/L, PO BOX 3105, WESTON CREEK 2611. PHONE: (06) 288 1677 FAX: (06) 288 2021

Please forward .... copies of **Wings Across the Sea**

.... Soft cover $19.95 .... Hard cover $29.95
(Please add $4.95 per item for postage)
($A30 or $A40 respectively for overseas seamail orders.)

(Please allow 16 days for delivery)

❒ I enclose my cheque/MO for $ .... payable to AEROSPACE PUBLICATIONS P/L or please charge to:

❒ Bankcard ❒ MasterCard ❒ Visa ❒ Amex

Card No: .... Expiry: ....

Ph: .... Signature: ....

NAME ....

ADDRESS ....

.... P/C ....

THE BIBLE OF AVIATION

AVAILABLE DIRECT FROM AUSTRALIAN AVIATION

The latest edition of the famous **Observers Book of Aircraft** lists a wealth of data on 140 different aircraft.

Included is information on country of origin, powerplant, performance, armament (where appropriate), weights, specifications and project status. Each fixed wing aircraft is illustrated by a high quality three view silhouette and the text is supported by over 250 photographs.

New types included are the A330/340, Boeing 777, Mriya, An-74, Beriev Mermaid, YF-23, F-22, B-2, F-117A, MiG-31, Yak-141, 747-400, Schweizer 330, MDH Explorer, Kamov Hokum, Eurocopter Tiger, Il-96-300, Tu-204 and dozens of other types not covered in previous editions.

TO: AEROSPACE PUBLICATIONS P/L, PO BOX 3105, WESTON CREEK 2611. PHONE: (06) 288 1677 FAX: (06) 288 2021

Please forward .... copies of Observers Aircraft @ $15.00 each
(post free in Aust, int'l orders add $A5.00)
(Please allow 16 days for delivery)

❒ I enclose my cheque/MO for $ .... payable to AEROSPACE PUBLICATIONS P/L or please charge to:

❒ Bankcard ❒ Mastercard ❒ Visa ❒ Amex

Card No: .... Expiry: ....

Ph: .... Signature: ....

NAME ....

ADDRESS ....

.... P/C ....